T0202777

Lecture Notes in Mathematics 2211

More information about this series at http://www.springer.com/series/304

Fondazione C.I.M.E., Firenze

C.I.M.E. stands for *Centro Internazionale Matematico Estivo*, that is, International Mathematical Summer Centre. Conceived in the early fifties, it was born in 1954 in Florence, Italy, and welcomed by the world mathematical community: it continues successfully, year for year, to this day.

Many mathematicians from all over the world have been involved in a way or another in C.I.M.E.'s activities over the years. The main purpose and mode of functioning of the Centre may be summarised as follows: every year, during the summer, sessions on different themes from pure and applied mathematics are offered by application to mathematicians from all countries. A Session is generally based on three or four main courses given by specialists of international renown, plus a certain number of seminars, and is held in an attractive rural location in Italy.

The aim of a C.I.M.E. session is to bring to the attention of younger researchers the origins, development, and perspectives of some very active branch of mathematical research. The topics of the courses are generally of international resonance. The full immersion atmosphere of the courses and the daily exchange among participants are thus an initiation to international collaboration in mathematical research.

C.I.M.E. Director (2002 – 2014)
Pietro Zecca
Dipartimento di Energetica "S. Stecco"
Università di Firenze
Via S. Marta, 3
50139 Florence
Italy
e-mail: zecca@unifi.it

C.I.M.E. Director (2015 –)
Elvira Mascolo
Dipartimento di Matematica "U. Dini"
Università di Firenze
viale G.B. Morgagni 67/A
50134 Florence
Italy
e-mail: mascolo@math.unifi.it

C.I.M.E. Secretary
Paolo Salani
Dipartimento di Matematica "U. Dini"
Università di Firenze
viale G.B. Morgagni 67/A
50134 Florence
Italy
e-mail: salani@math.unifi.it

CIME activity is carried out with the collaboration and financial support of INdAM (Istituto Nazionale di Alta Matematica)

For more information see CIME's homepage: **http://www.cime.unifi.it**

Alessio Figalli • Ireneo Peral • Enrico Valdinoci

Partial Differential Equations and Geometric Measure Theory

Cetraro, Italy 2014

Alberto Farina • Enrico Valdinoci
Editors

Authors
Alessio Figalli
Department of Mathematics
ETH Zurich
Zurich, Switzerland

Ireneo Peral
Departamento de Matemáticas
Universidad Autónoma de Madrid
Madrid, Spain

Enrico Valdinoci
Dipartimento di Matematica
Università di Milano
Milano, Italy

Editors
Alberto Farina
LAMFA
University of Picardie at Amiens
Amiens, France

Enrico Valdinoci
Dipartimento di Matematica
Università di Milano
Milano, Italy

ISSN 0075-8434 ISSN 1617-9692 (electronic)
Lecture Notes in Mathematics
C.I.M.E. Foundation Subseries
ISBN 978-3-319-74041-6 ISBN 978-3-319-74042-3 (eBook)
https://doi.org/10.1007/978-3-319-74042-3

Library of Congress Control Number: 2018935159

Mathematics Subject Classification (2010): 35J15, 26A33, 35R35

Printed on acid-free paper

This Springer imprint is published by the registered company Springer International Publishing AG part of Springer Nature.
The registered company address is: Gewerbestrasse 11, 6330 Cham, Switzerland

Introduction

This collection of notes consists of lectures given during the school "Partial Differential Equations and Geometric Measure Theory", held in Cetraro in the period June 2–7, 2014.

The lectures were given by Henry Berestycki (Centre d'analyse et de mathématique sociales, *Reaction-diffusion and propagation in non-homogeneous media*), László Székelyhidi (Universität Leipzig, *The h-principle, the Nash-Kuiper theorem and the Euler equations*), Alessio Figalli (University of Texas at Austin, *Monge-Ampère type equation and applications*), Frank Morgan (Williams College, *Geometric Measure Theory, Isoperimetric Problems, and Manifolds with Density*), and Ireneo Peral (Universidad Autónoma de Madrid, *Elliptic and parabolic equations related to growth models*).

The topic of the school included a variety of fields of research, in which a crucial role is played by the interplay of fine analytic techniques and deep geometric observations, thus combining the intuitive and geometric aspect of mathematics with the methods coming from analysis and variational methods.

The problems addressed in these fields of research are challenging and complex, and they often require the use of several refined techniques to overcome the major difficulties encountered. To make these methods available to a wide audience, the speakers of the courses were chosen among the world leading experts who contributed enormously to the advance of the research and who possess great communication skills.

The outcome of the event was great, in terms of exchange of scientific information and also at a personal level. The speakers and the participants were enthusiastic and we think that this CIME Course was very effective in promoting new scientific interactions.

The lectures by Henry Berestycki presented a series of modern results on reaction-diffusion and propagation in nonhomogeneous media. Not only these equations provide a great source of interesting and beautiful mathematical problems, but they also arise in a variety of models in biomathematics and the validity of the results obtained in the theory of partial differential equations can be directly confronted with experiments and historical data. A natural application of the methods discussed in the course is indeed the measurement of the diffusion of some diseases, or of the speed of invasion of some biological species, especially when the spreading possibility is enhanced by fast channels or roads of fast diffusion.

The course of László Székelyhidi focused on the h-principle, which is a concept introduced by Gromov. Intuitively a system of partial differential equations (or, more generally, of partial differential inequalities) is said to satisfy the h-principle if exactly the opposite of "rigidity" can be proven, i.e. if there is a wide abundance of solutions. This is often the case for underdetermined systems, but one observes occasionally the striking phenomenon that some (geometrically relevant) overdetermined systems of partial differential equations also obey the h-principle. The primary example is the highly counterintuitive Nash-Kuiper C^1 isometric embedding theorem: say, for every $\varepsilon > 0$ there is a C^1 isometric embedding of the standard 2-dimensional sphere in the 3-dimensional Euclidean ball of radius ε, namely one can bend without stretching or tearing a page of paper and make it as small as she or he wishes! (of course, the issue is how regular this embedding can be to violate the preservation of the Gaussian curvature).

The course gave a general introduction to the ideas behind the h-principle following the book of Eliashberg and Mischachev, also discussing the Smale's sphere eversion theorem. Related (but much more complicated) techniques can be used in the case of the incompressible Euler equations, giving the first example of continuous solutions which dissipate the kinetic energy. These solutions were conjectured to exist in 1949 by Lars Onsager, in connection with Kolmogorov's theory of turbulence and his "energy cascade". Indeed, the full conjecture of Onsager states the existence of dissipative Hölder solutions for every Hölder exponent strictly smaller than $\frac{1}{3}$ and the absence of dissipation for any Hölder solution with exponent strictly larger than $\frac{1}{3}$. The latter part of the conjecture has been proved by Eyink and Constantin-E-Titi in the 1990s, whereas the first is still open.

The course of Frank Morgan covered most of his book on Geometric Measure Theory. In particular, the syllabus of the course can be summarized as follows:

- Rectifiable sets,
- Rectifiable currents,
- Compactness theorem and existence,
- Regularity,
- Flat chains modulo ν, varifolds, and Almgren-minimal sets,
- Soap bubble clusters,
- Hexagonal honeycomb and Kelvin conjectures,
- Manifolds with density, the Log-Convex Density Conjecture, and Perelman's proof of the Poincaré Conjecture.

The course ended with a very enjoyable, fun, and instructive soap bubble activity, which showed us how mathematics is deeply related with concrete phenomena of nature and how some (at least at a first glance) "abstract" mathematical notions of measure theory, topology, and calculus of variations play a crucial role if we want to understand the complicated patterns that even very simple situations, such as a game for kids, may produce.

The slides of Frank Morgan's contribution were kindly made available by the author on the webpage

https://www.ma.utexas.edu/users/enrico/Morgan-notes-cetraro2014/

Frank Morgan has also posted several pictures of those days on the webpage

http://sites.williams.edu/Morgan/2014/06/01/grand-hotel-san-michele/

As for the courses of Alessio Figalli and Ireneo Peral, we are not going to spoil the surprise to the reader, who will find in the forthcoming pages a neat and complete exposition of the courses given by the authors themselves.

We also include here a contribution, dealing with the curious case of s-harmonic functions and with their great flexibility in terms of approximation, which in turn provides a number of specific applications.

We hope that this book can give at least a faint hint of the striking "live" experience of the school in Cetraro by trying to leave a permanent trace of this "once in a lifetime" event, and to strongly suggest to students and young researchers of all age to take part in the forthcoming CIME Courses.

We would like indeed to express our warmest gratitude to the CIME Foundation, to the CIME Directors Professor Pietro Zecca and Professor Elvira Mascolo, to the CIME Board Secretary Paolo Salani, and to the CIME staff for their invaluable help and support, and for making the environment in Cetraro so stimulating and enjoyable.

And, once again, thanks a lot to the speakers and to the participants for their communicative efforts and their contagious enthusiasm!

Acknowledgements: CIME activity is carried out with the collaboration and financial support of:

- INdAM (Istituto Nazionale di Alta Matematica)
- MIUR (Ministero dell'Istruzione, dell'Università e della Ricerca)
- ERC "EPSILON" Elliptic Pde's and Symmetry of Interfaces and Layers for Odd Nonlinearities

Amiens, France
Milano, Italy

Alberto Farina
Enrico Valdinoci

Contents

Global Existence for the Semigeostrophic Equations via Sobolev Estimates for Monge-Ampère

Alessio Figalli

1 The Semigeostrophic Equations

The semigeostrophic (in short, SG) equations are a simple model used in meteorology to describe large scale atmospheric flows. As explained for instance in [5, Section 2.2] (see also [14] for a more complete exposition), these equations can be derived from the 3-D incompressible Euler equations, with Boussinesq and hydrostatic approximations, subject to a strong Coriolis force. Since for large scale atmospheric flows the Coriolis force dominates the advection term, the flow is mostly bi-dimensional. For this reason, the study of the SG equations in 2-D or 3-D is pretty similar, and in order to simplify our presentation we focus here on the 2-dimensional periodic case.

1.1 The Classical SG System

The 2-dimensional SG system can be written as

$$\begin{cases} \partial_t \nabla p_t + (\boldsymbol{u}_t \cdot \nabla)\nabla p_t + \nabla^\perp p_t + \boldsymbol{u}_t = 0 & \text{in } [0,\infty)\times\mathbb{R}^2, \\ \operatorname{div} \boldsymbol{u}_t = 0 & \text{in } [0,\infty)\times\mathbb{R}^2, \\ p_0 = \bar{p} & \text{on } \mathbb{R}^2, \end{cases} \tag{1.1}$$

A. Figalli (✉)
Mathematics Department, The University of Texas at Austin, Austin, TX, USA
e-mail: figalli@math.utexas.edu

A. Farina, E. Valdinoci (eds.), *Partial Differential Equations and Geometric Measure Theory*, Lecture Notes in Mathematics 2211,
https://doi.org/10.1007/978-3-319-74042-3_1

where $\boldsymbol{u}_t = (\boldsymbol{u}_t^1, \boldsymbol{u}_t^2) : \mathbb{R}^2 \to \mathbb{R}^2$ and $p_t : \mathbb{R}^2 \to \mathbb{R}$ are time-dependent periodic[1] functions respectively corresponding to the velocity and the pressure.

In the above system the notation $\nabla^\perp p_t$ denotes the rotation of the vector ∇p_t by $\pi/2$, that is $\nabla^\perp p_t = (\partial_2 p_t, -\partial_1 p_t)$. Also, $(\boldsymbol{u}_t \cdot \nabla)$ denotes the operator $\boldsymbol{u}_t^1 \partial_1 + \boldsymbol{u}_t^2 \partial_2$. Hence, in components the first equation in (1.1) reads as

$$\partial_t \partial_1 p_t + \sum_{j=1,2} \boldsymbol{u}_t^j \partial_j \partial_1 p_t + \partial_2 p_t + \boldsymbol{u}_t^1 = 0,$$

$$\partial_t \partial_2 p_t + \sum_{j=1,2} \boldsymbol{u}_t^j \partial_j \partial_2 p_t - \partial_1 p_t + \boldsymbol{u}_t^2 = 0.$$

Notice that in (1.1) we have three equations (the two above, together with div $\boldsymbol{u}_t = 0$) for the three unknowns $(p_t, \boldsymbol{u}_t^1, \boldsymbol{u}_t^2)$. Also, while in many equations in fluid mechanics one usually prescribes the evolution of the velocity field $\boldsymbol{u}_t$ and p_t acts as a Lagrange multiplier for the incompressibility constraint, here we are prescribing the evolution of the gradient of p_t and $\boldsymbol{u}_t$ acts as a Lagrange multiplier in order to ensure that the vector field ∇p_t remains a gradient along the evolution.

As shown in [14], energetic considerations show that it is natural to assume that p_t is (-1)-convex, i.e., the function $P_t(x) := p_t(x) + |x|^2/2$ is convex on $\mathbb{R}^2$. Hence, noticing that

$$\nabla p_t = \nabla P_t - x, \qquad \partial_t \nabla p_t = \partial_t \nabla P_t, \qquad (\boldsymbol{u}_t \cdot \nabla) x = \boldsymbol{u}_t,$$

we are led to the following extended system for P_t:

$$\begin{cases} \partial_t \nabla P_t + (\boldsymbol{u}_t \cdot \nabla) \nabla P_t + (\nabla P_t - x)^\perp = 0 & \text{in } [0,\infty) \times \mathbb{R}^2, \\ \operatorname{div} \boldsymbol{u}_t = 0 & \text{in } [0,\infty) \times \mathbb{R}^2, \\ P_t \text{ convex} & \text{in } [0,\infty) \times \mathbb{R}^2, \\ P_0 = \bar{p} + |x|^2/2 & \text{on } \mathbb{R}^2, \end{cases} \tag{1.2}$$

with the "boundary conditions" that both $P_t - |x|^2/2$ and $\boldsymbol{u}_t$ are periodic.

The existence theory for this equation is extremely complicated, and so far nobody has been able to attack directly this equation. Instead, there is a way to find a "dual equation" to this system for which existence of solutions is much easier to obtain, and then one can use this "dual solution" to go back and construct a solution to the original system. This is the goal of the next sections.

[1] By "periodic" we shall always mean $\mathbb{Z}^2$-periodic.

1.2 An Evolution Equation for the Density Associated to P_t: The Dual SG System

Notice that ∇P_t can be seen a map from $\mathbb{R}^2$ to $\mathbb{R}^2$. Motivated by the fact that, in optimal transport theory, gradients of convex functions are optimal transport maps (see Theorem 2.1 below), it is natural to look at the image of the Lebesgue measure under this map and try to understand its behavior. Hence, denoting by dx denote Lebesgue measure on $\mathbb{R}^2$, we define the measure ρ_t as $(\nabla P_t)_\sharp dx$, that is, for any test function $\chi : \mathbb{R}^2 \to \mathbb{R}$,

$$\int_{\mathbb{R}^2} \chi(y)\, d\rho_t(y) := \int_{\mathbb{R}^2} \chi\big(\nabla P_t(x)\big)\, dx. \tag{1.3}$$

Before describing the properties of ρ_t, we make a simple observation that will be useful later.

Remark 1 Since $\nabla P_t - x$ is periodic, it is easy to check that the measure ρ_t is periodic on $\mathbb{R}^2$ and

$$\int_{[0,1]^2} d\rho_t = \int_{[0,1]^2} dx = 1,$$

so one can also identify it as a probability measure on the 2-dimensional torus $\mathbb{T}^2$.

Our goal now is to find an evolution equation for ρ_t. To this aim we take $\varphi \in C_c^\infty(\mathbb{R}^2)$ and we compute the time derivative of $\int \varphi\, d\rho_t$:

$$\begin{aligned}
\frac{d}{dt}\int \varphi\, d\rho_t &\overset{(1.3)}{=} \frac{d}{dt}\int \varphi\big(\nabla P_t\big)\, dx \\
&= \int \nabla\varphi\big(\nabla P_t\big)\cdot \partial_t \nabla P_t\, dx \\
&\overset{(1.2)}{=} -\int \nabla\varphi\big(\nabla P_t\big)\cdot (\boldsymbol{u}_t\cdot\nabla)\nabla P_t\, dx - \int \nabla\varphi\big(\nabla P_t\big)\cdot (\nabla P_t - x)^\perp\, dx \\
&= -\int \nabla\varphi\big(\nabla P_t\big)\cdot D^2 P_t\cdot \boldsymbol{u}_t\, dx - \int \nabla\varphi\big(\nabla P_t\big)\cdot (\nabla P_t - x)^\perp\, dx \\
&= -\int \nabla\big[\varphi\circ\nabla P_t\big]\cdot \boldsymbol{u}_t\, dx - \int \nabla\varphi\big(\nabla P_t\big)\cdot (\nabla P_t - x)^\perp\, dx \\
&= \int \big[\varphi\circ\nabla P_t\big]\operatorname{div}\boldsymbol{u}_t\, dx - \int \nabla\varphi\big(\nabla P_t\big)\cdot (\nabla P_t - x)^\perp\, dx \\
&\overset{(1.2)}{=} -\int \nabla\varphi\big(\nabla P_t\big)\cdot (\nabla P_t - x)^\perp\, dx,
\end{aligned} \tag{1.4}$$

where at the last step we used that $\operatorname{div}\boldsymbol{u}_t = 0$.

In order to continue in the computations we need to introduce the Legendre transform of P_t:

$$P_t^*(y) := \sup_{x\in\mathbb{R}^2} \{x \cdot y - P_t(x)\} \qquad \forall\, y \in \mathbb{R}^2.$$

Notice that the function P_t^* is also convex, being the supremum of linear functions. Also, at least formally, the gradient of P_t and P_t^* are inverse to each other[2]:

$$\nabla P_t(\nabla P_t^*(y)) = y, \qquad \nabla P_t^*(\nabla P_t(x)) = x. \tag{1.5}$$

Thanks to this fact we obtain that the last term in (1.4) is equal to

$$-\int \Big[\nabla\varphi \cdot (\nabla P_t^* - y)^{\perp}\Big] \circ \nabla P_t(x)\, dx,$$

which by (1.3) can also be rewritten as

$$-\int \nabla\varphi \cdot (\nabla P_t^* - y)^{\perp}\, d\rho_t(y).$$

Hence, if we set

$$\boldsymbol{U}_t(y) := \big(\nabla P_t^*(y) - y\big)^{\perp},$$

(1.4) and an integration by parts give

$$\frac{d}{dt}\int \varphi\, d\rho_t = -\int \varphi\, \mathrm{div}(\boldsymbol{U}_t\rho_t),$$

[2]The relation (1.5) is valid only at point where the gradients of P_t and P_t^* both exist. There is however a weaker way to formulate such a relation that is independent of any regularity assumption: define the sub-differential of a convex function $\phi : \mathbb{R}^n \to \mathbb{R}$ as

$$\partial\phi(x) := \{p \in \mathbb{R}^n \,:\, \phi(z) \geq \phi(x) + p\cdot(z-x) \quad \forall\, z \in \mathbb{R}^n\}.$$

Then, using the definition of P_t^* it is not difficult to check that

$$y \in \partial P_t(x) \quad \Leftrightarrow \quad x \in \partial P_t^*(y).$$

Noticing that $\partial\phi(x) = \{\nabla\phi(x)\}$ whenever ϕ is differentiable at x, the above relation reduces exactly to (1.5) at differentiability points.

and by the arbitrariness of φ we conclude that $\partial_t\rho_t + \mathrm{div}(\boldsymbol{U}_t\rho_t) = 0$. Thus we have shown that ρ_t satisfies the following *dual problem*:

$$\begin{cases} \partial_t\rho_t + \mathrm{div}(\boldsymbol{U}_t\rho_t) = 0 & \text{in } [0,\infty)\times\mathbb{R}^2, \\ \boldsymbol{U}_t(y) = \left(\nabla P_t^*(y) - y\right)^\perp & \text{in } [0,\infty)\times\mathbb{R}^2, \\ \rho_t = (\nabla P_t)_\sharp dx & \text{in } [0,\infty)\times\mathbb{R}^2, \\ P_0 = \bar{p} + |x|^2/2 & \text{on } \mathbb{R}^2. \end{cases} \tag{1.6}$$

2 Global Existence for the Dual SG System

The global existence of weak solutions to the dual problem (1.6) was obtained by Benamou and Brenier in [5]. The aim of this section is to review their result.

2.1 *Preliminaries on Transport Equations*

The system (1.6) is given by a transport equation for ρ_t where the vector field $\boldsymbol{U}_t$ is coupled to ρ_t via the relation $\rho_t = (\nabla P_t)_\sharp dx$. Notice that because $\boldsymbol{U}_t = (\boldsymbol{U}_t^1, \boldsymbol{U}_t^2)$ is the rotated gradient of the function $p_t^*(y) := P_t^*(y) - |y|^2/2$, it is divergence free: indeed,

$$\mathrm{div}\,\boldsymbol{U}_t = \partial_1\boldsymbol{U}_t^1 + \partial_2\boldsymbol{U}_t^2 = \partial_1\partial_2 p_t^* - \partial_2\partial_1 p_t^* = 0.$$

We now review some basic facts on linear transport equations with Lipschitz divergence-free vector fields. Since the dimension does not play any role, we work directly in $\mathbb{R}^n$.

Let $\boldsymbol{v}_t : \mathbb{R}^n \to \mathbb{R}^n$ be a time-dependent Lipschitz divergence-free vector field. Given a measure $\bar{\sigma}$ in $\mathbb{R}^n$, our goal is to solve the equation

$$\begin{cases} \partial_t\sigma_t + \mathrm{div}(\boldsymbol{v}_t\sigma_t) = 0 & \text{in } [0,\infty)\times\mathbb{R}^n, \\ \sigma_0 = \bar{\sigma} & \text{on } \mathbb{R}^n. \end{cases} \tag{2.1}$$

Notice that because $\boldsymbol{v}_t$ is divergence free, the above equation can also be rewritten as a standard transport equation:

$$\partial_t\sigma_t + \boldsymbol{v}_t\cdot\nabla\sigma_t = 0.$$

To find a solution, we first apply the classical Cauchy-Lipschitz theorem for ODEs in order to construct a flow for $\boldsymbol{v}_t$:

$$\begin{cases} \dot{Y}(t,y) = \boldsymbol{v}_t(Y(t,y)) \\ Y(0,y) = y, \end{cases} \tag{2.2}$$

and then we define $\sigma(t) := Y(t)_{\#}\bar{\sigma}$. Let us check that this definition provides a solution to (2.1): take $\varphi \in C_c^\infty(\mathbb{R}^n)$ and observe that

$$\begin{aligned} \frac{d}{dt}\int \varphi(y)\,d\sigma_t(y) &\overset{(1.3)}{=} \frac{d}{dt}\int \varphi(Y(t,y))\,d\bar{\sigma}(y) \\ &= \int \nabla\varphi(Y(t,y))\cdot \dot{Y}(t,y)\,d\bar{\sigma}(y) \\ &= \int \nabla\varphi(Y(t,y))\cdot \boldsymbol{v}_t(Y(t,y))\,d\bar{\sigma}(y) \\ &\overset{(1.3)}{=} \int \nabla\varphi(y)\cdot \boldsymbol{v}_t(y)\,d\sigma_t(y). \end{aligned}$$

By the arbitrariness of φ, this proves the validity of (2.1).

It is interesting to observe that the curve of measures $t \mapsto Y(t)_{\#}\bar{\sigma}$ is the unique solution of (2.1). A possible way to prove this is to consider σ_t an arbitrary solution of (2.1) and define $\hat{\sigma}_t := [Y(t)^{-1}]_{\#}\sigma_t$. Then a direct computation shows that

$$\frac{d}{dt}\int \varphi(y)\,d\hat{\sigma}_t(y) = 0 \qquad \forall \in C_c^\infty(\mathbb{R}^2), \tag{2.3}$$

therefore

$$[Y(t)^{-1}]_{\#}\sigma_t = \hat{\sigma}_t \overset{(2.3)}{=} \hat{\sigma}_0 = \bar{\sigma} \quad \Rightarrow \quad \sigma_t = Y(t)_{\#}\bar{\sigma},$$

as desired.

We also notice that, if we assume that $\bar{\sigma}$ is not just a measure but a function, then we can give a more explicit formula for σ_t. Indeed the fact that $\operatorname{div} \boldsymbol{v}_t = 0$ implies that $\det\nabla Y(t) = 1$,[3] and the classical change of variable formula gives

$$\int \varphi(y)\,d\sigma_t(y) \overset{(1.3)}{=} \int \varphi(Y(t,y))\,\sigma(y)\,dy \overset{z=Y(t,y)}{=} \int \varphi(z)\,\bar{\sigma}\big(Y(t)^{-1}(z)\big)\,dz.$$

[3]To show this, one differentiates (2.2) with respect to y and uses the classical identity

$$\frac{d}{d\varepsilon}_{|\varepsilon=0} \det(A+\varepsilon BA) = \operatorname{tr}(B)\,\det(A),$$

Since φ is arbitrary we deduce that σ_t is a function (and not just a measure) and it is given by the formula $\sigma_t = \bar{\sigma} \circ Y(t)^{-1}$, or equivalently

$$\sigma_t(Y(t,y)) = \bar{\sigma}(y) \qquad \forall y. \tag{2.5}$$

This implies in particular that any pointwise bound on $\bar{\sigma}$ is preserved in time, that is

$$\lambda \le \bar{\sigma} \le \Lambda \quad \Rightarrow \quad \lambda \le \sigma_t \le \Lambda \qquad \forall t. \tag{2.6}$$

2.2 Preliminaries on Optimal Transport

Let μ, ν denote two probability measures on $\mathbb{R}^n$. The optimal transport problem (with quadratic cost) consists in finding the "optimal" way to transport μ onto ν, given that the transportation cost to move a point from x to y is $|x-y|^2$. Hence one is naturally led to minimize

$$\int_{\mathbb{R}^n} |S(x) - x|^2 \, d\mu(x)$$

among all maps S which send μ onto ν, that is $S_{\#}\mu = \nu$. By a classical theorem of Brenier [9] existence and uniqueness of optimal maps always hold provided μ is absolutely continuous and both μ and ν have finite second moments. In addition, the optimality of the map is equivalent to the fact that T is the gradient of a convex function. This is summarized in the next theorem:

Theorem 2.1 *Let μ, ν be probability measures on $\mathbb{R}^n$ with $\mu = f\,dx$ and*

$$\int |x|^2 \, d\mu(x) + \int |y|^2 \, d\nu(y) < \infty.$$

Then:

(1) There exists a unique optimal transport map T.
(2) There exists a convex function $\phi : \mathbb{R}^n \to \mathbb{R}$ such that $T = \nabla\phi$.
(3) The fact that T is the gradient of a convex function is equivalent to its optimality. More precisely, if $\psi : \mathbb{R}^n \to \mathbb{R}$ is a convex function such that $\nabla\psi_{\#}\mu = \nu$ then

---

to get

$$\begin{cases} \frac{d}{dt}\big(\det \nabla Y(t,y)\big) = [\operatorname{div} \boldsymbol{v}_t(Y(t,y))]\big(\det \nabla Y(t,y)\big) = 0, \\ \det \nabla Y(0,y) = y. \end{cases} \tag{2.4}$$

$\nabla\psi$ is optimal and $T = \nabla\psi$. In addition, if $f > 0$ a.e. then $\psi = \phi + c$ for some additive constant $c \in \mathbb{R}$.

We now show the connection between optimal transport and the Monge-Ampère equation. Assume that both μ and ν are absolutely continuous, that is $\mu = f\,dx$ and $\nu = g\,dy$, let $\varphi \in C_c^\infty(\mathbb{R}^n)$, and suppose that $T = \nabla\phi$ is a smooth diffeomorphism. Then, using the definition of push-forward and the standard change of variable formula, we get

$$\int \varphi(T(x))f(x)\,dx \overset{(1.3)}{=} \int \varphi(y)\,g(y)\,dy \overset{y=T(x)}{=} \int \varphi(T(x))\,g(T(x))\,\left|\det\nabla T(x)\right|\,dx.$$

By the arbitrariness of φ, this gives the Jacobian equation

$$f(x) = g(T(x))\left|\det\nabla T(x))\right|,$$

that combined with the condition $T = \nabla\phi$ implies that ϕ solves the Monge-Ampère equation

$$\det(D^2\phi) = \frac{f}{g\circ\nabla\phi}. \tag{2.7}$$

Notice that the above computations are formal since we needed to assume a priori T to be smooth in order to write the above equation. Still, this fact is the starting point behind the regularity theory of optimal transport maps. We shall not enter into this but we refer to the survey paper [19] for more details.

Notice that in Sect. 1.2 we started from the Lebesgue measure on $\mathbb{R}^2$ and we sent it onto ρ_t using the gradient of the convex function P_t. If we could apply Theorem 2.1(3) above we would know that ∇P_t is the unique optimal map sending the Lebesgue measure onto ρ_t. However in our case we do not have probability measure but rather periodic measures on $\mathbb{R}^n$, hence Theorem 2.1 does not directly apply. However, since both the Lebesgue measure and ρ_t are probability measures on the torus (see Remark 1), we can apply [13] (see also [2, Theorem 2.1]) to obtain the following:

Theorem 2.2 *Let μ, ν be probability measures on $\mathbb{T}^2$, and assume that $\mu = f\,dx$ and that $f > 0$ a.e. Then there exists a unique (up to an additive constant) convex function $P : \mathbb{R}^2 \to \mathbb{R}$ such that $(\nabla P)_\sharp\mu = \nu$ and $P - |x|^2/2$ is periodic.*

2.3 *Dual SG vs. 2-D Euler*

Before entering into the proof of existence of solutions to (1.6), let us first make a parallel with the 2-dimensional Euler equations. Starting from the Euler system

$$\begin{cases} \partial_t \boldsymbol{u}_t + (\boldsymbol{u}_t\cdot\nabla)\boldsymbol{u}_t + \nabla p_t = 0 \\ \operatorname{div}\boldsymbol{u}_t = 0 \end{cases} \tag{2.8}$$

one can consider the curl of $\boldsymbol{u}_t$ given by $\omega_t := \partial_1 \boldsymbol{u}_t^2 - \partial_2 \boldsymbol{u}_t^1$. Then, by taking the curl of the first equation in (2.8), one finds that ω_t solves the equation

$$\begin{cases} \partial_t \omega_t + \operatorname{div}(\boldsymbol{u}_t\, \omega_t) = 0 \\ \omega_t = \operatorname{curl} \boldsymbol{u}_t \\ \operatorname{div} \boldsymbol{u}_t = 0. \end{cases} \tag{2.9}$$

Since $\boldsymbol{u}_t$ is divergence free, it follows that $\operatorname{curl} \boldsymbol{u}_t^\perp = 0$, hence $\boldsymbol{u}_t^\perp$ is the gradient of a function ψ_t, or equivalently $\boldsymbol{u}_t = -\nabla^\perp \psi_t$. Then, inserting this information inside the relation $\omega_t = \operatorname{curl} \boldsymbol{u}_t$ we deduce that $\omega_t = -\operatorname{curl} \nabla^\perp \psi_t = \Delta \psi_t$, and (2.9) rewrites as

$$\begin{cases} \partial_t \omega_t + \operatorname{div}(\boldsymbol{u}_t\, \omega_t) = 0 \\ \boldsymbol{u}_t = \nabla^\perp \psi_t \\ \Delta \psi_t = \omega_t \end{cases} \tag{2.10}$$

(see [16, Section 1.2.1] for more details).

If we now compare (2.10) and (1.6), we can notice that the two systems are very similar. More precisely, since the linearization of the determinant around the identity matrix gives the trace, we see that (1.6) is a nonlinear version of (2.10). This can be formalized in the following way (see [30, Section 6] for a rigorous result in this direction):

Exercise Assume that $(\rho_t^\varepsilon, P_t^{*,\varepsilon})_{\varepsilon>0}$ is a family of solutions to (1.6) with

$$\rho_t^\varepsilon = 1 + \varepsilon\, \omega_{\varepsilon t} + o(\varepsilon), \qquad P_t^{*,\varepsilon} = |y|^2/2 + \varepsilon\, \psi_{\varepsilon t} + o(\varepsilon),$$

for some couple of smooth functions (ω_t, ψ_t). Then (ω_t, ψ_t) solve (2.10).

2.4 Global Existence of Weak Solutions to (1.6)

In order to construct solutions to (1.6) one uses the following splitting method:

1. Given ρ_0, we construct the vector field $\boldsymbol{U}_0$ using the optimal transport map sending ρ_0 to dx, and we use (a regularization of) it to let ρ_0 evolve over a time interval $[0, \varepsilon]$.
2. Starting from ρ_ε, we construct $\boldsymbol{U}_\varepsilon$ as before and we use it to let ρ_ε evolve over the time interval $[\varepsilon, 2\varepsilon]$.
3. Iterating this procedure, we obtain an approximate solution on $[0, \infty)$, and letting $\varepsilon \to 0$ produces the desired solutions.

We now describe in detail this construction.

2.4.1 Construction of Approximate Solutions

Assume that $\rho_0 := (x + \nabla \bar{p})_{\#}dx$ satisfies

$$\lambda \leq \rho_0 \leq \Lambda \tag{2.11}$$

for some constants $0 \leq \lambda \leq \Lambda$.[4] Since ρ_0 is absolutely continuous, we can apply Theorem 2.2 in order to find a convex function P_0^* whose gradient sends ρ_0 to dx, and we define

$$\boldsymbol{U}_0(y) := \left(\nabla P_0^*(y) - y\right)^{\perp}.$$

The idea is to fix $\varepsilon > 0$ a small time step and to solve the transport equation in (1.6) over the time interval $[0, \varepsilon]$ but with $\boldsymbol{U}_0$ in place of $\boldsymbol{U}_t$, using the theory described in Sect. 2.1. However, since $\boldsymbol{U}_0$ is not smooth, we shall first regularize it.[5] For this reason with introduce a regularization parameter $\delta > 0$.[6]

Hence, we fix a smooth convolution kernel $\chi \in C_c^\infty(\mathbb{R}^2)$ and, for $t \in [0, \varepsilon]$, we define

$$\boldsymbol{U}_t^{\varepsilon,\delta}(y) := \boldsymbol{U}_0 * \chi_\delta(y) = \int_{\mathbb{R}^2} \boldsymbol{U}_0(z)\, \chi_\delta(y - z)\, dz, \qquad \chi_\delta(y) := \frac{1}{\delta^2}\, \chi\left(\frac{y}{\delta}\right).$$

Notice that $\boldsymbol{U}_t^{\varepsilon,\delta} \in C^\infty(\mathbb{R}^2)$ and $\operatorname{div} \boldsymbol{U}_t^{\varepsilon,\delta} = (\operatorname{div} \boldsymbol{U}_0) * \chi_\delta = 0$, hence we can apply the theory discussed in Sect. 2.1 in the following way: we denote by $Y^{\varepsilon,\delta}(t)$ the flow of $\boldsymbol{U}_t^{\varepsilon,\delta}$ on $[0, \varepsilon]$, that is

$$\begin{cases} \dot{Y}^{\varepsilon,\delta}(t, y) = \boldsymbol{U}_t^{\varepsilon,\delta}(Y^{\varepsilon,\delta}(t, y)) & \text{for } t \in [0, \varepsilon], \\ Y^{\varepsilon,\delta}(0, y) = y, \end{cases}$$

and define

$$\rho_t^{\varepsilon,\delta} := Y^{\varepsilon,\delta}(t)_{\#}\rho_0 \qquad \forall\, t \in [0, \varepsilon].$$

Since $\boldsymbol{U}_t^{\varepsilon,\delta}$ is divergence-free, it follows from (2.11) and (2.6) that

$$\lambda \leq \rho_t^{\varepsilon,\delta} \leq \Lambda \qquad \forall\, t \in [0, \varepsilon].$$

[4]In this proof the lower bound on ρ_0 is not crucial and this is why we are allowing for $\lambda = 0$ as a possible value. However, instead of just writing $\rho_0 \leq \Lambda$ we have decided to write (2.11) in order to emphasize that both the upper and the lower bound will be preserved along the flow.

[5]This regularization procedure is not strictly necessary, since in this situation one could also apply the theory of flows for divergence-free BV vector fields [1]. However, to keep the presentation elementary, we shall not use these advanced results.

[6]One could decide to choose $\delta = \varepsilon$ and to have only one small parameter. However, for clarity of the presentation, we prefer to keep these two parameter distinct.

We now "update" the vector field: we apply Theorem 2.2 to find a convex function $P_\varepsilon^{*,\varepsilon,\delta}$ whose gradient send $\rho_\varepsilon^{\varepsilon,\delta}$ to dx, we set

$$U_t^{\varepsilon,\delta}(y) := \left(\nabla P_\varepsilon^{*,\varepsilon,\delta} - y\right)^\perp * \chi_\delta(y) \qquad \forall t \in [\varepsilon, 2\varepsilon],$$

we consider $Y^{\varepsilon,\delta}(t)$ the flow of $U_t^{\varepsilon,\delta}$ on $[\varepsilon, 2\varepsilon]$,

$$\begin{cases} \dot{Y}^{\varepsilon,\delta}(t,y) = U_t^{\varepsilon,\delta}(Y^{\varepsilon,\delta}(t,y)) & \text{for } t \in [\varepsilon, 2\varepsilon], \\ Y^{\varepsilon,\delta}(\varepsilon, y) = y, \end{cases}$$

and we define

$$\rho_t^{\varepsilon,\delta} := Y^{\varepsilon,\delta}(t)_\# \rho_\varepsilon^{\varepsilon,\delta} \qquad \forall t \in [\varepsilon, 2\varepsilon].$$

This allows us to update again the vector field on the time interval $[2\varepsilon, 3\varepsilon]$ using the optimal map from $\rho_{2\varepsilon}^{\varepsilon,\delta}$ to dx, and so on. Iterating this procedure infinitely many times and defining

$$P_t^{*,\varepsilon,\delta} := P_{k\varepsilon}^{*,\varepsilon,\delta} \qquad \text{for } t \in [k\varepsilon, (k+1)\varepsilon),\ k \in \mathbb{N},$$

we construct a family of densities $\{\rho_t^{\varepsilon,\delta}\}_{t\geq 0}$ and vector fields $\{U_t^{\varepsilon,\delta}\}_{t\geq 0}$ satisfying

$$\begin{cases} \partial_t \rho_t^{\varepsilon,\delta} + \operatorname{div}(U_t^{\varepsilon,\delta} \rho_t^{\varepsilon,\delta}) = 0 & \text{in } [0,\infty) \times \mathbb{R}^2, \\ U_t^{\varepsilon,\delta} = \left(\nabla P_t^{*,\varepsilon,\delta} - y\right)^\perp * \chi_\delta & \text{in } [0,\infty) \times \mathbb{R}^2, \\ (\nabla P_t^{*,\varepsilon,\delta})_\# \rho_{k\varepsilon} = dx & \text{for } t \in [k\varepsilon, (k+1)\varepsilon), \\ \lambda \leq \rho_t^{\varepsilon,\delta} \leq \Lambda & \text{in } [0,\infty) \times \mathbb{R}^2, \\ \rho_0 = (x + \nabla \bar{p})_\# dx & \text{on } \mathbb{R}^2. \end{cases} \tag{2.12}$$

2.4.2 Taking the Limit in the Approximate System

Notice that, because $\nabla P_t^{*,\varepsilon,\delta}$ are optimal transport maps between probability densities on the torus, it is not difficult to show that

$$|\nabla P_t^{*,\varepsilon,\delta}(y) - y| \leq \operatorname{diam}(\mathbb{T}^2) = \frac{\sqrt{2}}{2} \qquad \forall y \in \mathbb{R}^2 \tag{2.13}$$

(see [2, Theorem 2.1]), which implies that the vector fields $U_t^{\varepsilon,\delta}$ are uniformly bounded. Hence, given an arbitrary sequence $\varepsilon, \delta \to 0$, up to extracting a subsequence we can find densities ρ_t and a bounded vector field $\boldsymbol{U}_t$ such that

$$\rho_t^{\varepsilon,\delta} \rightharpoonup^* \rho_t \qquad \text{in } L^\infty_{\text{loc}}\big([0,\infty)\times\mathbb{R}^2\big),$$

$$\boldsymbol{U}_t^{\varepsilon,\delta} \rightharpoonup^* \boldsymbol{U}_t \qquad \text{in } L^\infty_{\text{loc}}\big([0,\infty)\times\mathbb{R}^2;\mathbb{R}^2\big).$$

In particular, since $\lambda \leq \rho_t^{\varepsilon,\delta} \leq \Lambda$, it follows immediately that ρ_t satisfies

$$\lambda \leq \rho_t \leq \Lambda \qquad \text{for a.e.} t \geq 0.$$

- **Step 1: Find the Limit of $\boldsymbol{U}_t^{\varepsilon,\delta}\rho_t^{\varepsilon,\delta}$.** In order to take the limit into (2.12), the most difficult term to deal with is the product $\boldsymbol{U}_t^{\varepsilon,\delta}\rho_t^{\varepsilon,\delta}$ inside the divergence, since in general it is not true that under weak convergence this would converge to $\boldsymbol{U}_t\rho_t$. However in this case we can exploit extra regularity.

  More precisely, since both $\rho_t^{\varepsilon,\delta}$ and $\boldsymbol{U}_t^{\varepsilon,\delta}$ are uniformly bounded, we see that for any smooth function $\psi:\mathbb{R}^2\to\mathbb{R}$ it holds

$$\int \operatorname{div}(\boldsymbol{U}_t^{\varepsilon,\delta}\rho_t^{\varepsilon,\delta})\psi\,dy = -\int \boldsymbol{U}_t^{\varepsilon,\delta}\cdot\nabla\psi\,\rho_t^{\varepsilon,\delta}\,dy \leq C\|\psi\|_{W^{1,1}(\mathbb{R}^2)}.$$

  This means that $\operatorname{div}(\boldsymbol{U}_t^{\varepsilon,\delta}\rho_t^{\varepsilon,\delta})$ belongs to $[W^{1,1}(\mathbb{R}^2)]^*$ (the dual space of $W^{1,1}(\mathbb{R}^2)$) uniformly in time, which implies that

$$\partial_t\rho_t^{\varepsilon,\delta} = -\operatorname{div}(\boldsymbol{U}_t^{\varepsilon,\delta}\rho_t^{\varepsilon,\delta}) \in L^\infty\big([0,\infty),[W^{1,1}(\mathbb{R}^2)]^*\big) \subset L^p_{\text{loc}}\big([0,\infty),[W^{1,q}_{\text{loc}}(\mathbb{R}^2)]^*\big)$$

  for any $p\in[1,\infty]$ and $q\geq 1$ (here we used that $W^{1,q}_{\text{loc}}(\mathbb{R}^2)\subset W^{1,1}_{\text{loc}}(\mathbb{R}^2)$ to get the opposite inclusion between dual spaces). Combining this regularity in time with the bound

$$\rho_t^{\varepsilon,\delta} \in L^\infty\big((0,\infty),L^\infty(\mathbb{R}^2)\big) \subset L^p_{\text{loc}}\big([0,\infty),L^p_{\text{loc}}(\mathbb{R}^2)\big),$$

  by the Aubin-Lions Lemma (see for instance [32]) we deduce that

$$\text{the family}\,\rho_t^{\varepsilon,\delta}\,\text{is precompact in}\, L^p_{\text{loc}}\big([0,\infty),[W^{s,q}_{\text{loc}}(\mathbb{R}^2)]^*\big) \text{ for any } p<\infty, q>1, s>0,$$

  hence

$$\rho_t^{\varepsilon,\delta} \to \rho_t \qquad \text{in } L^p_{\text{loc}}\big([0,\infty),[W^{s,q}_{\text{loc}}(\mathbb{R}^2)]^*\big) \text{ for any } p<\infty, q>1, s>0. \tag{2.14}$$

  In order to exploit this strong compactness we need to gain some regularity in space on $\boldsymbol{U}_t^{\varepsilon,\delta}$.

To this aim, observe that being $P_t^{*,\varepsilon,\delta}$ smooth convex functions, one can easily get an a-priori bound on their Hessian: since for a non-negative symmetric matrix the norm is controlled by the trace, using the divergence theorem and the uniform local Lipschitzianity of $P_t^{*,\varepsilon,\delta}$ (see (2.13)) we get

$$\int_{B_R} \|D^2 P_t^{*,\varepsilon,\delta}\|\, dy \leq \int_{B_R} \Delta P_t^{*,\varepsilon,\delta}\, dy \leq \int_{\partial B_R} |\nabla P_t^{*,\varepsilon,\delta}|\, dy \leq C_R \qquad \forall R > 0. \tag{2.15}$$

By fractional Sobolev embeddings (see [7, Chapter 6]) we deduce that, uniformly with respect to ε and δ,

$$\boldsymbol{U}_t^{\varepsilon,\delta} \in L^\infty\big((0,\infty), W^{1,1}_{\mathrm{loc}}(\mathbb{R}^2)\big) \subset L^\infty\big((0,\infty), W^{s,q}_{\mathrm{loc}}(\mathbb{R}^2)\big)$$

for all $s \in (0,1)$ and $1 \leq q < \frac{2}{1+s}$. In particular, choosing for instance $s = 1/2$ and $q = 5/4$ we deduce that

$$\boldsymbol{U}_t^{\varepsilon,\delta} \rightharpoonup^* \boldsymbol{U}_t \qquad \text{in } L^\infty\big((0,\infty), W^{1/2,5/4}_{\mathrm{loc}}(\mathbb{R}^2)\big),$$

that combined with (2.14) with $s = 1/2$ and $q = 5/4$ implies that

$$\boldsymbol{U}_t^{\varepsilon,\delta} \rho_t^{\varepsilon,\delta} \rightharpoonup^* \boldsymbol{U}_t \rho_t \qquad \text{in } L^\infty_{\mathrm{loc}}\big((0,\infty) \times \mathbb{R}^2; \mathbb{R}^2\big).$$

This allows us to pass to the limit in the transport equation in the distributional sense and deduce that

$$\partial_t \rho_t + \operatorname{div}(\boldsymbol{U}_t \rho_t) = 0 \qquad \text{in } (0,\infty) \times \mathbb{R}^2.$$

- **Step 2: show that $\boldsymbol{U}_t = (\nabla P_t^* - y)^\perp$.** To conclude the proof we need so show that if P_t^* is the convex function sending ρ_t onto dx (see Theorem 2.2) then

$$\boldsymbol{U}_t = (\nabla P_t^* - y)^\perp \qquad \text{for a.e. } t \geq 0. \tag{2.16}$$

To this aim notice that (2.14) implies that, up to a subsequence,

$$\rho_t^{\varepsilon,\delta} \to \rho_t \qquad \text{in } [W^{s,q}_{\mathrm{loc}}(\mathbb{R}^2)]^* \textit{ for a.e. } t \geq 0,$$

hence, being $\rho_t^{\varepsilon,\delta}$ uniformly bounded in L^∞, we also deduce that

$$\rho_t^{\varepsilon,\delta} \rightharpoonup^* \rho_t \qquad \text{in } L^\infty(\mathbb{R}^2) \textit{ for a.e. } t \geq 0.$$

By stability of optimal transport maps (see for instance [34, Corollary 5.23]) it follows that[7]

$$\nabla P_t^{*,\varepsilon,\delta} \to \nabla P_t^* \qquad \text{in } L^1_{\mathrm{loc}}(\mathbb{R}^2) \textit{ for a.e. } t \geq 0.$$

Recalling that $U_t^{\varepsilon,\delta} = (\nabla P_t^{*,\varepsilon,\delta} - y)^\perp * \chi_\delta$ we deduce that

$$U_t^{\varepsilon,\delta} \to (\nabla P_t^* - y)^\perp \qquad \text{in } L^1_{\mathrm{loc}}(\mathbb{R}^2) \textit{ for a.e. } t \geq 0,$$

which shows the validity of (2.16) and concludes the proof of the existence of weak solutions.

Notice that, as a consequence of (2.13), the uniform bound

$$|\nabla P_t^*(y) - y| \leq \frac{\sqrt{2}}{2} \tag{2.17}$$

holds. This will be useful in the sequel.

3 Back from Dual SG to SG

In the previous section we have constructed a weak solution (ρ_t, P_t^*) to the dual system (1.6). Also, we have seen that if $\rho_0 := (x + \nabla \bar{p})_\# dx$ satisfies $\lambda \leq \rho_0 \leq \Lambda$ then these bounds are propagated in time, that is

$$\lambda \leq \rho_t \leq \Lambda \qquad \text{for a.e. } t \geq 0. \tag{3.1}$$

In this section we shall assume that $\lambda > 0$.

3.1 A Formula for (p_t, u_t)

We want to use the solution (ρ_t, P_t^*) to construct a couple $(p_t, \boldsymbol{u}_t)$ solving the original SG systems (1.2).

[7] Actually, if one assumes that $\lambda > 0$ (that is the densities $\rho_t^{\varepsilon,\delta}$ are uniformly bounded away from zero and infinity) then the convergence of $\nabla P_t^{*,\varepsilon,\delta}$ to ∇P_t^* holds even in $W^{1,1}_{\mathrm{loc}}(\mathbb{R}^2)$, see [18].

3.1.1 Construction of p_t

Recalling the procedure used to go from p_t to P_t^* (adding $|x|^2/2$ to p_t and taking a Legendre transform), it is easy to perform the inverse construction and define p_t from P_t^*: indeed, if we define[8]

$$P_t(x) := \sup_{y\in\mathbb{R}^2}\{\, y\cdot x - P_t^*(y)\} \tag{3.2}$$

and set

$$p_t(x) := P_t(x) - \frac{|x|^2}{2}, \tag{3.3}$$

thanks to the periodicity of $P_t^* - |\,y|^2/2$ it is easy to check that p_t is periodic too. Hence, constructing p_t given P_t^* is relatively simple.

3.1.2 Construction of $\boldsymbol{u}_t$

More complicated is the formula for $\boldsymbol{u}_t$. Let us start from (1.2). From the first equation and the fact that D^2P_t is a symmetric matrix, we get

$$D^2P_t\cdot \boldsymbol{u}_t = -\partial_t\nabla P_t - (\nabla P_t - x)^{\perp}. \tag{3.4}$$

In order to invert D^2P_t, we differentiate (1.5) with respect to x to find that

$$D^2P_t^*(\nabla P_t)\,D^2P_t = \mathrm{Id}, \tag{3.5}$$

while differentiating (1.5) with respect to t gives

$$[\partial_t\nabla P_t^*](\nabla P_t) + D^2P_t^*(\nabla P_t)\cdot\partial_t\nabla P_t = 0 \tag{3.6}$$

Hence, thanks to (3.5), multiplying both sides of (3.4) by $D^2P_t^*(\nabla P_t)$ we get

$$\boldsymbol{u}_t = -D^2P_t^*(\nabla P_t)\cdot\partial_t\nabla P_t - D^2P_t^*(\nabla P_t)\cdot(\nabla P_t - x)^{\perp},$$

that combined with (3.6) gives

$$\boldsymbol{u}_t = [\partial_t\nabla P_t^*](\nabla P_t) - D^2P_t^*(\nabla P_t)\cdot(\nabla P_t - x)^{\perp}. \tag{3.7}$$

---

[8]Recall that the Legendre transform is an involution on convex functions, that is, if $\phi:\mathbb{R}^n\to\mathbb{R}$ is convex then $(\phi^*)^* = \phi$.

Hence we have found an expression of $\boldsymbol{u}_t$ in terms of derivatives of P_t^* and its Legendre transform. However the problem is to give a meaning to such terms.

First of all one may ask what is $D^2P_t^*(\nabla P_t)$. Notice that being P_t^* a convex function, a priori $D^2P_t^*$ is only a matrix-valued measure, thus it is not clear what $D^2P_t^*(\nabla P_t)$ means. However, if we remember that $(\nabla P_t^*)_\#\rho_t = dx$, it follows by the discussion in Sect. 2.2 (see in particular (2.7)) that

$$\det(D^2P_t^*) = \rho_t. \tag{3.8}$$

Hence, recalling (3.1) we deduce that

$$\lambda \le \det(D^2P_t^*) \le \Lambda.$$

As we shall see in Sect. 4.1 below, this condition implies that

$$P_t^* \in W^{2,1+\gamma}_{\mathrm{loc}}(\mathbb{R}^2) \quad \text{for some } \gamma = \gamma(n,\lambda,\Lambda) > 0. \tag{3.9}$$

We claim that this estimate allows us to give a meaning to $D^2P_t^*(\nabla P_t)$ and prove that

$$D^2P_t^*(\nabla P_t) \in L^\infty\big((0,\infty), L^{1+\gamma}_{\mathrm{loc}}(\mathbb{R}^2)\big).$$

Indeed, since $(\nabla P_t^*)_\#\rho_t = dx$, it follows from (1.5) that $(\nabla P_t)_\# dx = \rho_t$. Also, since p_t is periodic we see that $\nabla P_t = x + \nabla p_t$ is a bounded perturbation of the identity, hence there exists $C > 0$ such that

$$\nabla P_t(B_R) \subset B_{R+C} \qquad \forall R > 0.$$

These two facts imply that

$$\begin{aligned}
\int_{B_R} \|D^2P_t^*(\nabla P_t)\|^{1+\gamma}\,dx &\overset{(1.3)}{=} \int_{\nabla P_t(B_R)} \|D^2P_t^*\|^{1+\gamma}\rho_t(y)\,dy \\
&\overset{(3.1)}{\le} \Lambda \int_{B_{R+C}} \|D^2P_t^*\|^{1+\gamma}\,dy \overset{(3.9)}{<} \infty
\end{aligned}$$

for all $R > 0$, hence

$$D^2P_t^*(\nabla P_t) \in L^\infty\big((0,\infty), L^{1+\gamma}_{\mathrm{loc}}(\mathbb{R}^2)\big).$$

Recalling that $(\nabla P_t - x)^\perp$ is globally bounded (see (2.17)), we deduce that

$$D^2P_t^*(\nabla P_t)\cdot(\nabla P_t - x)^\perp \in L^\infty\big((0,\infty), L^{1+\gamma}_{\mathrm{loc}}(\mathbb{R}^2)\big),$$

so the last term in (3.7) is a well defined function.

Concerning the term $[\partial_t \nabla P_t^*](\nabla P_t)$, as explained in Sect. 4.2 one can show that

$$\partial_t \nabla P_t^* \in L^{1+\kappa}_{\mathrm{loc}} \qquad \text{for } \kappa = \frac{\gamma}{2+\gamma} > 0, \tag{3.10}$$

and arguing as above one deduces that

$$[\partial_t \nabla P_t^*](\nabla P_t) \in L^\infty\big((0,\infty), L^{1+\kappa}_{\mathrm{loc}}(\mathbb{R}^2)\big).$$

In conclusion we have seen that, using (3.9) and (3.10), the formula (3.7) defines a function $\boldsymbol{u}_t \in L^\infty\big((0,\infty), L^{1+\kappa}_{\mathrm{loc}}(\mathbb{R}^2)\big)$, which is easily seen to be periodic.

Hence, modulo the validity of (3.9) and (3.10), we have constructed a couple of functions $(p_t, \boldsymbol{u}_t)$ that we expect to solve (1.1). In the next section we shall see that the functions $(p_t, \boldsymbol{u}_t)$ defined in (3.3) and (3.7) are indeed solutions of (1.1), and then in Sect. 4 we will prove both (3.9) and (3.10).

3.2 (p_t, u_t) Solves the Semigeostrophic System

In order to prove that $(p_t, \boldsymbol{u}_t)$ is a distributional solution of (1.1) we need to find some suitable test functions to use in (1.6).

More precisely, we first write (1.6) in distributional form:

$$\int \int_{\mathbb{T}^2} \Big\{ \partial_t \varphi_t(x) + \nabla \varphi_t(x) \cdot \boldsymbol{U}_t(x) \Big\}\, \rho_t(x)\, dx\, dt\, dx = 0 \tag{3.11}$$

for every $\varphi \in W^{1,1}((0,\infty) \times \mathbb{R}^2)$ periodic in the space variable.

We now take $\phi \in C_c^\infty((0,\infty) \times \mathbb{R}^2)$ a function periodic in space, and we consider the test function $\varphi : (0,\infty) \times \mathbb{R}^2 \to \mathbb{R}^2$ defined as

$$\varphi_t(y) := J\,(y - \nabla P_t^*(y))\, \phi_t\big(\nabla P_t^*(y)\big), \tag{3.12}$$

where J denotes the matrix corresponding to the rotation by $\pi/2$, that is

$$J := \begin{pmatrix} 0 & -1 \\ 1 & 0 \end{pmatrix}.$$

Notice that $Jv = -v^\perp$ for any $v \in \mathbb{R}^2$, hence φ_t can be equivalently written as

$$\varphi_t := (\nabla P_t^* - y)^\perp\, \phi_t(\nabla P_t^*)$$

We compute the derivatives of φ:

$$\begin{cases} \partial_t \varphi_t = [\partial_t \nabla P_t^*]^\perp \phi_t(\nabla P_t^*) + (\nabla P_t^* - y)^\perp \partial_t \phi_t(\nabla P_t^*) \\ \qquad +(y - \nabla P_t^*)^\perp \big[\nabla \phi_t(\nabla P_t^*) \cdot \partial_t \nabla P_t^*\big], \\ \nabla \varphi_t = J\big(Id - D^2 P_t^*\big)\, \phi_t(\nabla P_t^*) + (\nabla P_t^* - y)^\perp \otimes \big(\nabla \phi_t(\nabla P_t^*) \cdot D^2 P_t^*\big). \end{cases} \tag{3.13}$$

Since $\boldsymbol{U}_t = (\nabla P_t^* - y)^\perp$ and $(\nabla P_t^*)_\sharp \rho_t = dx$, recalling (1.5) we can use (3.13) to rewrite (3.11) as

$$\begin{aligned} 0 &= \int_0^\infty \int_{\mathbb{T}^2} \Big\{\partial_t \varphi_t + \nabla \varphi_t \cdot \boldsymbol{U}_t\Big\} \rho_t(y)\, dy\, dt \\ &= \int_0^\infty \int_{\mathbb{T}^2} \Big\{[\partial_t \nabla P_t^*]^\perp (\nabla P_t)\, \phi_t + (x - \nabla P_t)^\perp\, \partial_t \phi_t \\ &\quad + (x - \nabla P_t)^\perp \big[\nabla \phi_t \cdot [\partial_t \nabla P_t^*](\nabla P_t)\big] \\ &\quad + \Big[J\big(Id - D^2 P_t^*(\nabla P_t)\big)\, \phi_t + (x - \nabla P_t)^\perp \\ &\quad \otimes \big(\nabla \phi_t \cdot D^2 P_t^*(\nabla P_t)\big)\Big](x - \nabla P_t)^\perp\Big\}\, dx\, dt. \end{aligned}$$

Taking into account the formula (3.7) for $\boldsymbol{u}_t$, after rearranging the terms the above expression yields

$$0 = \int_0^\infty \int_{\mathbb{T}^2} \Big\{-\nabla^\perp p_t \left(\partial_t \phi_t + \boldsymbol{u}_t \cdot \nabla \phi_t\right) + \left(-\nabla p_t + \boldsymbol{u}_t^\perp\right) \phi_t\Big\}\, dx\, dt,$$

hence $(p_t, \boldsymbol{u}_t)$ solve the first equation in (1.1). The fact that $\boldsymbol{u}_t$ is divergence free is obtained in a similar way, using the test function

$$\varphi_t := \phi(t)\, \psi(\nabla P_t{}^*)$$

where $\phi \in C_c^\infty((0, \infty))$, and $\psi \in C_c^\infty(\mathbb{R}^2)$ is periodic.

This shows that $(p_t, \boldsymbol{u}_t)$ is a distributional solution of (1.1), and we obtain the following result (see [2, Theorem 1.2]):

Theorem 3.1 *Let $\bar{p} : \mathbb{R}^2 \to \mathbb{R}$ be a periodic function such that $\bar{p}(x) + |x|^2/2$ is convex, and assume that the measure $\bar{\rho} := (Id + \nabla \bar{p})_\sharp dx$ is absolutely continuous with respect to the Lebesgue measure and satisfies $0 < \lambda \leq \bar{\rho} \leq \Lambda$.*

Let (ρ_t, P_t^*) be a solution of (1.6) starting from $\bar{\rho}$ satisfying $0 < \lambda \leq \rho_t \leq \Lambda$, and let P_t be as in (3.2). Then the couple $(p_t, \boldsymbol{u}_t)$ defined in (3.3) and (3.7) is a distributional solution of (1.1).

Although the vector field $\boldsymbol{u}_t$ provided by the previous theorem is only $L^{1+\kappa}_{\mathrm{loc}}$, in [2] the authors showed how to associate to it a measure-preserving Lagrangian flow. In particular, this allowed them to recover (in the particular case of the 2-dimensional periodic setting) the result of Cullen and Feldman [15] on the existence of Lagrangian solutions to the SG equations in physical space (see also [22, 23] for some recent results on the existence of weak Lagrangian solutions).

As shown in [3], the above result can also be generalized to bounded convex domain $\Omega \subset \mathbb{R}^3$. However this extension presents several additional difficulties. Indeed, first of all in 3-dimensions the SG system becomes much less symmetric compared to its 2-dimensional counterpart, because the action of Coriolis force regards only the first and the second space components. Moreover, even considering regular initial data and velocities, several arguments in the proofs require a finer regularization scheme. Still, under suitable assumptions on the initial data one can prove global existence of distributional solutions (see [3] for more details).

4 Regularity Estimates for the Monge-Ampère Equation

The aim of this section is to give a proof of the key estimates (3.9) and (3.10) used in the previous section to obtain global existence of distributional solutions to (1.1).

We shall first prove (3.9), and then show how (3.9) combined with (1.6) yields (3.10).

4.1 Sobolev Regularity for Monge-Ampère: Proof of (3.9)

Our goal is to show that, given $0 < \lambda \le \Lambda$, solutions to

$$\begin{cases} \lambda \le \det(D^2\phi) \le \Lambda \\ \phi \text{ convex} \\ \phi - |x|^2/2 \text{ periodic} \end{cases} \tag{4.1}$$

belong to $W^{2,1+\gamma}_{\mathrm{loc}}$ for some $\gamma > 0$ [17, 20, 31]. This result is valid in any dimension and restricting to dimension 2 would not simplify the proof. Also, since we want to prove an a-priori estimate on solutions to (4.1), one can assume that ϕ is smooth. Hence, we shall assume that $\phi : \mathbb{R}^n \to \mathbb{R}$ is a C^2 solution of (4.1) and we will show that

$$\int_{[0,1]^n} \|D^2\phi\|^{1+\gamma} \le C,$$

for some constant C depending only on n, λ, Λ. (From now on, any constant which depends only on n, λ, Λ will be called *universal*.)

We shall mainly follow the arguments in [17], except for Step 2 in Sect. 4.1.4 which is inspired by Schmidt [31].

4.1.1 Sections and Normalized Solutions

An important role in the regularity theory of Monge-Ampère is played by the sections of the function ϕ: given $x \in \mathbb{R}^n$ and $t > 0$, we define the section centered at x with height t as

$$S(x,t) := \big\{ y \in \Omega \,:\, u(y) \le u(x) + \nabla u(x) \cdot (y - x) + t \big\}. \tag{4.2}$$

Moreover, given $\tau > 0$, we use the notation $\tau S(x, p, t)$ to denote the dilation of $S(x, p, t)$ by a factor τ with respect to x, that is

$$\tau S(x,t) := \left\{ y \in \mathbb{R}^n \,:\, x + \frac{y - x}{\tau} \in S(x,t) \right\}. \tag{4.3}$$

Notice that, because $\phi - |x|^2/2$ is periodic, ϕ has quadratic growth at infinity. In particular its sections $S(x,t)$ are all bounded.

We say that an open bounded convex set $Z \subset \mathbb{R}^n$ is *normalized* if

$$B(0,1) \subset Z \subset B(0,n).$$

By John's Lemma [28], for every open bounded convex set there exists an (invertible) orientation preserving affine transformation $T : \mathbb{R}^n \to \mathbb{R}^n$ such that $T(Z)$ is normalized.

Notice that in the sequel we are not going to notationally distinguish between an affine transformation and its linear part, since it will always be clear to what we are referring to. In particular, we will use the notation

$$\|T\| := \sup_{|v|=1} |Av|, \qquad Tx = Ax + b.$$

One useful property which we will use is the following identity: if we denote by T^* the adjoint of T, then

$$\|T^* T\| = \|T^*\| \|T\|. \tag{4.4}$$

(This can be easily proved using the polar decomposition of matrices.)

Given a section $S(x,t)$, we can consider T an affine transformation which normalizes $S(x,t)$ and define the function

$$v(z) := (\det T)^{2/n} \Big[u(T^{-1}z) - u(x) - \nabla u(x) \cdot (T^{-1}z - x) - t \Big]. \tag{4.5}$$

Then it is immediate to check that v solves

$$\begin{cases} \lambda \le \det D^2 v \le \Lambda & \text{in } Z, \\ v = 0 & \text{on } \partial Z, \end{cases} \tag{4.6}$$

with $Z := T(S(x, t))$ normalized. We are going to call v a *normalized solution.*

As shown in [11] and [27] (see also [26, Chapter 3]), sections of solution of (4.1) satisfy strong geometric properties. We briefly recall here the ones that we are going to use[9]:

Proposition 4.1 *Let ϕ be a solution of* (4.1). *Then the following properties hold:*

(i) There exists a universal constant $\beta \in (0, 1)$ such that

$$\frac{1}{2}S(x, t) \subset S(x, t/2) \subset \beta S(x, t) \qquad \forall x \in \mathbb{R}^n,\ t > 0.$$

(ii) There exists a universal constant $\theta > 1$ such that

$$S(x, t) \cap S(y, t) \neq \emptyset \quad \Rightarrow \quad S(y, t) \subset S(x, \theta t) \qquad \forall x, y \in \mathbb{R}^n,\ t > 0.$$

(iii) There exists a universal constant $K > 1$ such that such that

$$\frac{t^{n/2}}{K} \le |S(x, t)| \le K\, t^{n/2} \qquad \forall x \in \mathbb{R}^n,\ t > 0.$$

(iv) $\bigcap_{t>0} S(x, t) = \{x\}$.

4.1.2 A Preliminary Estimate for Normalized Solutions

In this section we consider v a solution of (4.6) with $B_r \subset Z \subset B_R$ and we prove the following classical lemma due to Alexandrov:

Lemma 4.2 *Assume that v a solution of* (4.6) *with $B_r \subset Z \subset B_R$ for some universal radii $0 < r \le R$. There exist two universal constants $c_1, c_2 > 0$ such that*

$$c_1 \le \left|\inf_Z v\right| \le c_2, \tag{4.7}$$

Proof Set

$$g_-(z) := \frac{\lambda^{1/n}}{4}\big(|z|^2 - r^2\big).$$

[9] Usually all these properties are stated for small sections (say, when $t \le \rho$ for some universal ρ). However, since in our case ϕ is a global solution which has quadratic growth at infinity, it is immediate to check that all the properties are true when t is large.

We claim that $v \leq g_-$. Indeed, if not, let $c > 0$ be the smallest constant such that $v - c \leq g_-$ in Z, so that

$$v - c \leq g_- \quad \text{in } Z, \qquad v(\bar{z}) - c = g_-(\bar{z}) \quad \text{for some } \bar{z} \in \overline{Z}.$$

Notice that because $g_- \leq 0$ on ∂Z (since $B_r \subset Z$), the contact point $\bar{z}$ must be in the interior of Z. Hence, since the functions $g_- - (v - c)$ attains a local minimum at $\bar{z}$, its Hessian at $\bar{z}$ is non-negative definite, thus

$$D^2 g_-(\bar{z}) \geq D^2(v - c)(\bar{z}) = D^2 v(\bar{z}) \geq 0$$

which implies that

$$\frac{\lambda}{2^n} = \det\big(D^2 g_-(\bar{z})\big) \geq \det\big(D^2 v(\bar{z})\big) \geq \lambda,$$

a contradiction.

Set now

$$g_+(z) := \Lambda^{1/n}\big(|z|^2 - R^2\big).$$

A completely analogous argument based on the fact that $\det(D^2 g_+) = 2^n \Lambda$ and $g_+ \geq 0$ on ∂Z (since $Z \subset B_R$) shows that $g_+ \leq v$ in Z.

This proves that

$$g_+ \leq v \leq g_- \qquad \text{in } Z,$$

and the result follows.

4.1.3 Two Key Estimates on the Size of the Hessian

The following two lemmas are at the core of the proof of the $W^{2,1+\gamma}_{\mathrm{loc}}$ regularity. The first lemma estimates the L^1-size of $\|D^2\phi\|$ on a section $S(x,t)$, while the second one says that on a large fraction of points in $S(x,t)$ the value of $\|D^2\phi\|$ is comparable to its average.

Lemma 4.3 *Fix $x \in \mathbb{R}^n$, $t > 0$, and let T be the affine map which normalizes $S(x,t)$. Then there exists a positive universal constant C_1 such that*

$$\fint_{S(x,t)} \|D^2\phi\| \leq C_1 \frac{\|T\|\,\|T^*\|}{(\det T)^{2/n}} \tag{4.8}$$

Proof Consider the function $v : \mathbb{R}^n \to \mathbb{R}$ defined as in (4.5), and notice that

$$D^2 v(z) = (\det T)^{2/n} \left[(T^{-1})^* D^2 \phi(T^{-1} z) T^{-1} \right], \tag{4.9}$$

and

$$\begin{cases} \lambda \le \det D^2 v \le \Lambda & \text{in } T(S(x, 2t)), \\ v = \text{const.} & \text{on } \partial\big(T(S(x, 2t))\big). \end{cases} \tag{4.10}$$

Although the convex set $T(S(x, 2t))$ is not normalized in the sense defined before, it is almost so: indeed, since T normalizes $S(x, t)$, we have that

$$B_1 \subset T(S(x, 2t)). \tag{4.11}$$

Also, because

$$|S(x, 2t)| \le K(2t)^{n/2} \le 2^{n/2} K^2 |S(x, t)|$$

(by Proposition 4.1(iii)) and $T(S(x, t))$ is normalized, it follows that

$$|T(S(x, 2t))| \le 2^{n/2} K^2 |T(S(x, t))| \le 2^{n/2} K^2 |B_n| =: C_0,$$

where C_0 is universal. Since $T(S(x, 2t))$ is convex, the above estimate on its volume combined with (4.11) implies that

$$B_1 \subset T(S(x, 2t)) \subset B_R. \tag{4.12}$$

for some universal radius R. Hence, it follows from (4.10) and Lemma 4.2 that

$$\operatorname{osc}_{T(S(x,2t))} v \le c', \tag{4.13}$$

with c' universal.

Since v is convex, the size of its gradient is controlled by its oscillation in a slightly larger domain (see for instance [26, Lemma 3.2.1]), thus it follows from Proposition 4.1(i) and (4.13) that

$$\sup_{T(S(x,t))} |\nabla v| \le \sup_{\beta T(S(x,2t))} |\nabla v| \le \frac{\operatorname{osc}_{T(S(x,2t))} v}{\operatorname{dist}\Big(\beta T(S(x, 2t)),\ \partial\big(T(S(x, 2t))\big)\Big)} \le c'' \tag{4.14}$$

for some universal constant c''. Moreover, since $T(S(x, t))$ is a normalized convex set, it holds

$$|T(S(x, t))| \ge c_n \qquad \mathcal{H}^{n-1}\big(\partial T(S(x, t))\big) \le C_n, \tag{4.15}$$

where $c_n, C_n > 0$ are dimensional constants. Finally, since $D^2v(y)$ is non-negative definite (by the convexity of v) its norm is controlled by its trace, that is

$$\|D^2v(z)\| \leq \Delta v(z). \tag{4.16}$$

Thus, combining all these informations together we get

$$\begin{aligned}
\fint_{T(S(x,t))} \|D^2v(z)\|\,dz &\overset{(4.16)}{\leq} \fint_{T(S(x,t))} \Delta v(z)\,dz \\
&= \frac{1}{|T(S(x,t))|}\int_{\partial T(S(x,t))} \nabla v(z)\cdot \nu\, d\mathcal{H}^{n-1}(z) \\
&\overset{(4.15)}{\leq} \frac{C_n}{c_n}\Big(\sup_{T(S(x,t))}|\nabla v|\Big) \overset{(4.14)}{\leq} C_1,
\end{aligned} \tag{4.17}$$

that together with (4.9) gives

$$\begin{aligned}
\fint_{S(x,t)} \|D^2\phi(y)\|\,dy &= \frac{1}{(\det T)^{2/n}}\fint_{S(x,t)} \|T^*D^2v(Ty)T\|\,dy \\
&\leq \frac{\|T^*\|\|T\|}{(\det T)^{2/n}}\fint_{T(S(x,t))} \|D^2v(z)\|\,dz \leq C_1\frac{\|T^*\|\|T\|}{(\det T)^{2/n}},
\end{aligned}$$

concluding the proof. □

Lemma 4.4 *Fix $x \in \mathbb{R}^n$, $t > 0$, and let T be the affine map which normalizes $S(x,t)$. Then there exists a universal positive constant c_1 and a Borel set $A(x,t) \subset S(x,t)$, such that*

$$\frac{|A(x,t)\cap S(x,t)|}{|S(x,t)|} \geq \frac{1}{2} \tag{4.18}$$

and

$$\|D^2\phi(y)\| \geq c_1\frac{\|T\|\|T^*\|}{(\det T)^{2/n}} \qquad \forall\, y \in A(x,t). \tag{4.19}$$

Proof Let $v : \mathbb{R}^n \to \mathbb{R}$ defined as in (4.5), and recall that

$$\fint_{T(S(x,t))} \|D^2v(z)\|\,dz \leq C_1$$

for some universal constant C_1 (see (4.17)). Set

$$E := \{z \in T(S(x,t)) \;:\; \|D^2v(z)\| \geq 2C_1\}.$$

Then

$$2C_1\frac{|E|}{|T(S(x,t))|} \le \frac{1}{|T(S(x,t))|}\int_E \|D^2v(z)\|\,dz \le \fint_{T(S(x,t))} \|D^2v(z)\|\,dz \le C_1,$$

which implies that

$$|E| \le \frac{1}{2}|T(S(x,t))|.$$

Define $F := T(S(x,t)) \setminus E$ and notice that

$$\frac{|F|}{|T(S(x,t))|} \ge \frac{1}{2} \tag{4.20}$$

and (by (4.6) and the definition of E)

$$\begin{cases} \|D^2v\| \le 2C_1 & \text{inside } F, \\ \det(D^2v) \ge \lambda. \end{cases}$$

If we denote by $\alpha_1 \le \ldots \le \alpha_n$ the eigenvalues of D^2v, the first information tells us that $\alpha_n \le 2C_1$, while the second one that $\prod_i \alpha_i \ge \lambda$, from which it follows that

$$\alpha_1 \ge \frac{\lambda}{\prod_{i=2}^n \alpha_i} \ge \frac{\lambda}{(2C_1)^{n-1}} =: c_1,$$

therefore

$$c_1\mathrm{Id} \le D^2v \le 2C_1\mathrm{Id} \qquad \text{inside } F. \tag{4.21}$$

Recalling the definition of v (see (4.5)) this implies that

$$D^2\phi(y) = \frac{T^*D^2v(Ty)T}{(\det T)^{2/n}} \ge c_1\frac{TT^*}{(\det T)^{2/n}} \qquad \forall y \in A := T^{-1}(F),$$

so in particular

$$\|D^2\phi(y)\| \ge c_1\frac{\|T\|\|T^*\|}{(\det T)^{2/n}} \qquad \forall y \in A.$$

Finally, thanks to (4.20) we get

$$\frac{|A|}{|S(x,t)|} = \frac{|T(A)|}{|T(S(x,t))|} = \frac{|F|}{|T(S(x,t))|} \ge \frac{1}{2},$$

concluding the proof. □

4.1.4 Harmonic Analysis Related to Sections and the $W^{2,1+\gamma}_{\mathrm{loc}}$ Regularity

In this section we show how Lemmas 4.3 and 4.4 can be combined to obtain the desired result. Since the covering argument is slightly technical and may hide the ideas behind the proof, we prefer to give a formal argument and refer to the papers [17, 20, 31] for more details (see also [24]).

The basic idea behind the proof is that we can think of a section $S(x, t)$ as a "ball of radius t centered at x", and the properties stated in Proposition 4.1 ensure that sections are suitable objects to do harmonic analysis. Indeed it is possible to show that a Vitali Covering Lemma holds in this context (see for instance [20]), and that many standard quantities in harmonic analysis still enjoy all the properties that we are used to have in $\mathbb{R}^n$.

For instance, to $\|D^2\phi\|$ we can associated a "maximal function" using the sections:

$$\mathcal{M}(x) := \sup_{t>0} \fint_{S(x,t)} \|D^2\phi(y)\|\, dy \qquad \forall\, x \in \mathbb{R}^n.$$

Noticing that $D^2\phi$ is periodic, in order to deal with sets of finite volume we shall see both $D^2\phi$ and $\mathcal{M}$ as functions on the torus $\mathbb{T}^n$. In the same way, also the sections will be seen as subsets of $\mathbb{T}^n$ by considering the canonical projection $\pi : \mathbb{R}^n \to \mathbb{T}^n$.

The fact that sections behave like usual balls allows us to obtain the validity of a classical fact in harmonic analysis, that is that the L^1 norm of $\|D^2\phi\|$ on a super level sets $\{\|D^2\phi\| \geq \sigma\}$ is controlled by the measure where $\mathcal{M}$ is above σ (up to a universal constant). More precisely, by applying [33, Chapter 1, Section 4, Theorem 2] and [33, Chapter 1, Section 8.14], we deduce that the following holds: there exist universal constants $K, \sigma_0 > 0$ such that, for any $\sigma \geq \sigma_0$,

$$\int_{\{\|D^2\phi\|\geq\sigma\}} \|D^2\phi(y)\|\, dy \leq K\sigma \left|\left\{\mathcal{M} \geq \tfrac{\sigma}{K}\right\}\right|. \tag{4.22}$$

Our goal is to combine this estimate with Lemmas 4.3 and 4.4 to show that $D^2\phi \in W^{2,1+\gamma}_{\mathrm{loc}}$.

- **Step 1: Replace** $\mathcal{M}$ with $\|D^2\phi\|$ **in the Right Hand Side of** (4.22)**.** As we shall see, this is the step where we use Lemmas 4.3 and 4.4.

  Fix $\sigma \geq \sigma_0$. By the definition of $\mathcal{M}$, for any $x \in \{\mathcal{M} \geq \sigma/K\}$ we can find a section $S(x, t_x)$ such that

$$\fint_{S(x,t_x)} \|D^2\phi(y)\|\, dy \geq \frac{\sigma}{2K}. \tag{4.23}$$

Consider the family of sections $\{S(x,t_x)\}_{x\in\{\mathcal{M}\geq\sigma/K\}}$ constructed in this way, and extract a subfamily $\{S(x_i,t_{x_i})\}_{i\in I}$ such that

$$\left\{\mathcal{M}\geq\tfrac{\sigma}{K}\right\}\subset\bigcup_{i\in I}S(x_i,t_{x_i}) \tag{4.24}$$

and the sections $\{S(x_i,t_{x_i})\}_{i\in I}$ have bounded overlapping, that is,

$$\forall z\in\mathbb{T}^n\quad \#\{\,i\in I\,:\,z\in S(x_i,t_{x_i})\}\leq N \tag{4.25}$$

for some $N\in\mathbb{N}$ universal.[10]

Then, Lemmas 4.3 and 4.4 applied to the sections $S(x_i,t_{x_i})$ yield sets $A(x_i,t_{x_i})\subset S(x_i,t_{x_i})$ such that

$$\frac{|A(x_i,t_{x_i})|}{|S(x_i,t_{x_i})|}\geq\frac{1}{2},\qquad \#\{\,i\in I\,:\,z\in A(x_i,t_{x_i})\}\leq N\quad\forall z\in\mathbb{T}^n, \tag{4.26}$$

(the finite overlapping property is an immediate consequence of (4.25)), and

$$\frac{\sigma}{2K}\overset{(4.23)}{\leq}\fint_{S(x_i,t_{x_i})}\|D^2\phi(y)\|\,dy\leq C_1\frac{\|T_i\|\|T_i^*\|}{(\det T_i)^{2/n}}\leq\frac{C_1}{c_1}\|D^2\phi(y)\|\qquad\forall y\in A(x_i,t_{x_i}) \tag{4.27}$$

(here T_i denotes the affine map which normalizes $S(x_i,t_{x_i})$). Thanks to these facts we deduce that

$$\begin{aligned}\left|\left\{\mathcal{M}\geq\tfrac{\sigma}{K}\right\}\right| &\overset{(4.24)}{\leq}\sum_{i\in I}|S(x_i,t_{x_i})|\overset{(4.25)}{\leq}2\sum_{i\in I}|A(x_i,t_{x_i})|\\ &\overset{(4.27)}{\leq}2\sum_{i\in I}\left|A(x_i,t_{x_i})\cap\left\{\|D^2\phi\|\geq\tfrac{c_1\sigma}{2KC_1}\right\}\right|\\ &\overset{(4.26)}{\leq}2N\left|\left\{\|D^2\phi\|\geq\tfrac{c_1\sigma}{2KC_1}\right\}\right|.\end{aligned}$$

Hence, if we set $K_1:=\max\{2NK,2KC_1/c_1\}$, this allows us to replace $\mathcal{M}$ with $\|D^2\phi\|$ in the right hand side of (4.22) and get

$$\int_{\{\|D^2\phi\|\geq\sigma\}}\|D^2\phi(y)\|\,dy\leq K_1\sigma\left|\left\{\|D^2\phi\|\geq\tfrac{\sigma}{K_1}\right\}\right|\qquad\forall\sigma\geq\sigma_0. \tag{4.28}$$

[10] It is actually unknown whether, given a family of sections, one can extract a subfamily with finite overlapping. Here we are assuming that this can be done just to make the presentation simpler. However, there are at least to ways to circumvent this issue: either one slightly reduces t_{x_i} by a factor $(1-\varepsilon)$ with $\varepsilon>0$ so that the finite overlapping property holds (see [12, Lemma 1] and how this is applied in [17]), or one shrink t_{x_i} by a universal factor $\eta<1$ and then the sections can be made disjoint (see [20, 24]).

- **Step 2: A Gehring-Type Lemma.** Equation (4.28) is a sort of reverse Chebyshev's inequality for $\|D^2\phi\|$. We now show how this allows us to obtain higher integrability of $\|D^2\phi\|$.

  Set $g(s) := |\{\|D^2\phi\| \geq s\}|$. By the layer-cake formula we have

$$\int_{\{\|D^2\phi\|\geq\sigma\}} \|D^2\phi(y)\|\,dy = \sigma|\{\|D^2\phi\| \geq \sigma\}| + \int_\sigma^\infty |\{\|D^2\phi\| \geq s\}|\,ds$$
$$= g(\sigma)\sigma + \int_\sigma^\infty g(s)ds, \tag{4.29}$$

hence (4.28) implies that

$$\int_\sigma^\infty g(s)ds \leq K_1\sigma\, g\big(\tfrac{\sigma}{K_1}\big) \qquad \forall\, \sigma \geq \sigma_0. \tag{4.30}$$

Also, noticing that $g(\sigma) \leq |\mathbb{T}^n| = 1$, again by the layer-cake formula we get

$$\int_{\mathbb{T}^n} \|D^2\phi(y)\|^{1+\gamma}\,dy = (1+\gamma)\int_0^\infty \sigma^\gamma g(\sigma)\,d\sigma \leq \sigma_0^{1+\gamma} + (1+\gamma)\int_{\sigma_0}^\infty \sigma^\gamma g(\sigma)\,d\sigma.$$

Hence, to prove that $\|D^2\phi\| \in L^{1+\gamma}(\mathbb{T}^n)$ we have to show that

$$\int_{\sigma_0}^\infty \sigma^\gamma g(\sigma)\,d\sigma < \infty \tag{4.31}$$

for some $\gamma > 0$.

To this aim, performing an integrations by parts and using that $s \mapsto g(s)$ in non-increasing, we see that

$$\begin{aligned}
\int_{\sigma_0}^\infty \sigma^\gamma g(\sigma)\,d\sigma &= -\int_{\sigma_0}^\infty \sigma^\gamma \frac{d}{d\sigma}\left(\int_\sigma^\infty g(s)\,ds\right)d\sigma \\
&= \sigma_0^\gamma \int_{\sigma_0}^\infty g(s)\,ds + \gamma\int_{\sigma_0}^\infty \sigma^{\gamma-1}\left(\int_\sigma^\infty g(s)\,ds\right)d\sigma \\
&\overset{(4.30)}{\leq} \sigma_0^\gamma \int_{\sigma_0}^\infty g(s)\,ds + K_1\gamma\int_{\sigma_0}^\infty \sigma^\gamma g\big(\tfrac{\sigma}{K_1}\big)\,d\sigma \\
&\leq \sigma_0^\gamma \int_{\sigma_0}^\infty g(s)\,ds + K_1\gamma\, g\big(\tfrac{\sigma_0}{K_1}\big)\int_{\sigma_0}^{K_1\sigma_0} \sigma^\gamma\,d\sigma \\
&\quad + K_1\gamma\int_{K_1\sigma_0}^\infty \sigma^\gamma g\big(\tfrac{\sigma}{K_1}\big)\,d\sigma
\end{aligned}$$

$$\overset{\tau=\sigma/K_1}{=} \sigma_0^\gamma \int_{\sigma_0}^\infty g(s)\,ds + K_1 \frac{\gamma}{\gamma+1} (K_1\sigma_0)^{\gamma+1} g\big(\tfrac{\sigma_0}{K_1}\big)$$
$$+K_1^{2+\gamma} \gamma \int_{\sigma_0}^\infty \tau^\gamma g(\tau)\,d\tau.$$

Hence, recalling that $g \le 1$, we can choose $\gamma > 0$ small enough so that $K_1^{2+\gamma}\gamma \le 1/2$ and notice that

$$\sigma_0^\gamma \int_{\sigma_0}^\infty g(s)\,ds \overset{(4.29)}{\le} \int_{\{\|D^2\phi\|\ge\sigma\}} \|D^2\phi(y)\|\,dy \le \int_{\mathbb{T}^n} \|D^2\phi(y)\|\,dy < \infty$$

(to get the finiteness of $\|D^2\phi\|_{L^1(\mathbb{T}^n)}$ simply apply (4.8) with t large enough so that $S(x,t) \supset [0,1]^n$) to obtain that

$$\int_{\sigma_0}^\infty \sigma^\gamma g(\sigma)\,d\sigma \le 2\int_{\mathbb{T}^n} \|D^2\phi(y)\|\,dy + 2K_1 \frac{\gamma}{\gamma+1} (K_1\sigma_0)^{\gamma+1} < \infty.$$

This shows the validity of (4.31) and concludes the proof of the $W^{2,1+\gamma}_{\mathrm{loc}}$ regularity of ϕ.

4.2 *Regularity for Time-Dependent Solutions of Monge-Ampère: Proof of (3.10)*

To deal with the term $\partial_t \nabla P_t^*$, we shall use an idea of Loeper [29, Theorem 5.1] to combine (3.9) and (1.6) and prove the following:

Theorem 4.5 *There exists a universal constant C such that, for almost every $t \ge 0$,*

$$\int_{\mathbb{T}^2} \rho_t |\partial_t \nabla P_t^*|^{1+\kappa} \le C, \qquad \kappa := \frac{\gamma}{2+\gamma}. \tag{4.32}$$

Notice that, since $\rho_t \ge \lambda > 0$ (see (3.1)), (4.32) implies immediately (3.10).

Proof In order to justify the following computations one needs to perform a careful regularization argument. Here we show just the formal computations, referring to [2, Section 3] for more details.

We begin by differentiating in time the relation (3.8) to get

$$\sum_{i,j=1}^2 M_{ij}(D^2P_t^*)\,\partial_t\partial_{ij}P_t^* = \partial_t\rho_t,$$

where $M_{ij}(A) := \frac{\partial \det(A)}{\partial A_{ij}}$ is the cofactor matrix of A. Taking into account (1.6) and the well-known divergence-free property of the cofactor matrix

$$\sum_{i=1}^{2} \partial_i \big[M_{ij}(D^2 P_t{}^*) \big] = 0, \qquad j = 1, 2,$$

(see for instance [21, Chapter 8.1.4.b] for a proof), we can rewrite the above equation as

$$\sum_{i,j=1}^{2} \partial_i \big(M_{ij}(D^2 P_t^*)\, \partial_t \partial_j P_t^* \big) = -\mathrm{div}(\boldsymbol{U}_t \rho_t).$$

Then, recalling that for invertible matrices the cofactor matrix $M(A)$ is equal to $\det(A)\, A^{-1}$, using again the relation (3.8) we get

$$\mathrm{div}\big(\rho_t (D^2 P_t^*)^{-1} \partial_t \nabla P_t^* \big) = -\mathrm{div}(\rho_t \boldsymbol{U}_t). \tag{4.33}$$

We now multiply (4.33) by $\partial_t P_t^*$ and integrate by parts to obtain[11]

$$\begin{aligned} \int_{\mathbb{T}^2} \rho_t |(D^2 P_t^*)^{-1/2} \partial_t \nabla P_t^*|^2 \, dx &= \int_{\mathbb{T}^2} \rho_t \, \partial_t \nabla P_t^* \cdot (D^2 P_t^*)^{-1} \partial_t \nabla P_t^* \, dx \\ &= - \int_{\mathbb{T}^2} \rho_t \, \partial_t \nabla P_t^* \cdot \boldsymbol{U}_t \, dx. \end{aligned} \tag{4.34}$$

From Cauchy-Schwartz inequality, the right-hand side of (4.34) can be estimated as

$$\begin{aligned} &- \int_{\mathbb{T}^2} \rho_t \, \partial_t \nabla P_t^* \cdot (D^2 P_t^*)^{-1/2} (D^2 P_t^*)^{1/2} \boldsymbol{U}_t \, dx \\ &\quad \le \left(\int_{\mathbb{T}^2} \rho_t |(D^2 P_t^*)^{-1/2} \partial_t \nabla P_t^*|^2 \, dx \right)^{1/2} \left(\int_{\mathbb{T}^2} \rho_t |(D^2 P_t^*)^{1/2} \boldsymbol{U}_t|^2 \, dx \right)^{1/2}, \end{aligned} \tag{4.35}$$

hence (4.34) and (4.35) give

$$\int_{\mathbb{T}^2} \rho_t |(D^2 P_t^*)^{-1/2} \partial_t \nabla P_t^*|^2 \, dx \le \int_{\mathbb{T}^2} \rho_t |(D^2 P_t^*)^{1/2} \boldsymbol{U}_t|^2 \, dx. \tag{4.36}$$

[11] Since the matrix $D^2 P_t{}^*$ is positive definite, both its square root and the square root of its inverse are well-defined.

We now observe that

$$\int_{\mathbb{T}^2} \rho_t |(D^2 P_t^*)^{1/2} U_t|^2 \, dx = \int_{\mathbb{T}^2} \rho_t U_t \cdot D^2 P_t^* U_t \, dx \le \sup_{\mathbb{T}^2} \left(\rho_t |U_t|^2\right) \int_{\mathbb{T}^2} \|D^2 P_t^*\| \, dx. \tag{4.37}$$

Hence, recalling that U_t and ρ_t are bounded and noticing that $\int_{\mathbb{T}^2} \|D^2 P_t^*\| \, dx < \infty$,[12] it follows from (4.36) and (4.37) that

$$\int_{\mathbb{T}^2} \rho_t |(D^2 P_t^*)^{-1/2} \partial_t \nabla P_t^*|^2 \, dx \le C. \tag{4.38}$$

Thus, applying Hölder's inequality and noticing that $\frac{1+\kappa}{1-\kappa} = 1 + \gamma$, we get

$$\begin{aligned} \int_{\mathbb{T}^2} \rho_t |\partial_t \nabla P_t^*|^{1+\kappa} \, dx &\le \int_{\mathbb{T}^2} \left(\sqrt{\rho_t} |(D^2 P_t^*)^{-1/2} \partial_t \nabla P_t^*|\right)^{1+\kappa} \left(\frac{\|(D^2 P_t^*)^{1/2}\|}{\sqrt{\rho_t}}\right)^{1+\kappa} dx \\ &\le \left[\int_{\mathbb{T}^2} \rho_t |(D^2 P_t^*)^{-1/2} \partial_t \nabla P_t^*|^2 \, dx\right]^{(1+\kappa)/2} \\ &\quad \times \left[\int_{\mathbb{T}^2} \left(\frac{\|D^2 P_t^*\|}{\rho_t}\right)^{\frac{1+\kappa}{1-\kappa}} dx\right]^{(1-\kappa)/2} \\ &\overset{(4.38)+(3.1)}{\le} \left(\frac{C}{\lambda}\right)^{(1+\kappa)/2} \left[\int_{\mathbb{T}^2} \|D^2 P_t^*\|^{1+\gamma} \, dx\right]^{(1-\kappa)/2} \overset{(3.9)}{\le} \bar{C}, \end{aligned}$$

which proves (4.32). □

5 Short-Time Existence and Uniqueness of Smooth Solutions for Dual SG

In this section we discuss the results of Loeper in [30] concerning the short-time existence and uniqueness of smooth solutions for the dual SG system (1.6). As we have seen in the previous sections there is a strict correspondence between solutions of (1.6) and solutions of the original SG system (1.2), hence these results can be easily read back in the original framework.

We shall prove that if ρ_0 is Hölder continuous then there exists a unique Hölder solution (1.6) on some time interval $[0, T]$, where T depends only on the bounds on ρ_0. Using higher regularity theory for elliptic equations, it is not difficult to check that if ρ_0 is more regular (say, $C^{k,\alpha}$ for some $k \ge 0$ and $\alpha \in (0, 1)$), then the solution that we constructed enjoys the same regularity.

[12] This obviously follows by (3.9), but a direct proof can be given arguing as for (2.15).

The following result is contained in [30, Theorem 3.3, Corollary 3.4, Theorem 4.1]:

Theorem 5.1 *Assume that*

$$0 < \lambda \le \rho_0 \le \Lambda \quad \textit{and} \quad \rho_0 \in C^{0,\alpha}(\mathbb{T}^2) \tag{5.1}$$

for some $\alpha \in (0,1)$. *Then there exists* $T > 0$, *depending only on* λ, Λ, $\|\rho_0\|_{C^{0,\alpha}(\mathbb{T}^2)}$, *such that* (1.6) *has a unique solution* (ρ_t, P_t^*) *on* $[0,T]$ *satisfying*

$$0 < \lambda \le \rho_t \le \Lambda, \qquad \rho_t \in L^\infty\big([0,T], C^{0,\alpha}(\mathbb{T}^2)\big), \quad P_t^* \in L^\infty\big([0,T], C^{2,\alpha}(\mathbb{T}^2)\big). \tag{5.2}$$

We first discuss the existence part and then we deal with uniqueness.

5.1 Short-Time Existence of Smooth Solutions

The proof of existence given in [30, Theorem 3.3] is based on a fixed point argument. Here we give a different proof, more in the spirit of the argument used in Sect. 2.4.

Set $K_0 := 2\|\rho_0\|_{C^{0,\alpha}(\mathbb{T}^2)}$, let $T > 0$ small (to be fixed later), let $j \in \mathbb{N}$, and exactly as in Sect. 2.4.1 construct a family of approximate solutions by "freezing" the vector fields over time intervals of length $\frac{T}{j}$. More precisely, for $t \in \big[0, \frac{T}{j}\big]$ we define $P_t^{*,j}$ as the unique map whose gradient sends ρ_0 to dx (see Theorem 2.2), and we set

$$\boldsymbol{U}_t^j := \big(\nabla P_t^{*,j}(y) - y\big)^\perp \qquad \forall\, t \in \big[0, \tfrac{T}{j}\big]. \tag{5.3}$$

Notice that, by Caffarelli's regularity theory for the Monge-Ampère equation [10] we have

$$\|D^2 P_t^{*,j}\|_{C^{0,\alpha}(\mathbb{T}^2)} \le K_1 = K_1(K_0, \lambda, \Lambda) \qquad \forall\, t \in [0, T/t],$$

hence

$$\|\nabla \boldsymbol{U}_t^j\|_{L^\infty(\mathbb{T}^2)} \le K_2 := 1 + K_1. \tag{5.4}$$

We now consider the flow of $\boldsymbol{U}_t^j$ over the time interval $\big[0, \frac{T}{j}\big]$,

$$\begin{cases} \dot{Y}^j(t,y) = \boldsymbol{U}_t^j(Y^j(t,y)) & \text{for } t \in \big[0, \tfrac{T}{j}\big], \\ Y^j(0,y) = y, \end{cases} \tag{5.5}$$

and define

$$\rho_t^j := Y^j(t)_\# \rho_0 \qquad \forall\, t \in \left[0, \tfrac{T}{j}\right].$$

Recall that, since U_t^j is divergence free, ρ^j can also be written as

$$\rho_t^j = \rho_0 \circ Y^j(t)^{-1} \tag{5.6}$$

(see (2.5)). Recalling (5.1), this implies in particular that $\lambda \leq \rho_t^j \leq \Lambda$.

We now differentiate (5.5) with respect to y to get

$$\begin{cases} \frac{d}{dt}\nabla Y^j(t,y) = \Big(\nabla U_t^j(Y^j(t,y))\Big)\nabla Y^j(t,y), \\ \nabla Y^j(0,y) = \mathrm{Id}, \end{cases}$$

so (5.4) yields

$$\begin{cases} \frac{d}{dt}\|\nabla Y^j(t,y)\| \leq K_2\|\nabla Y^j(t,y)\|, \\ \|\nabla Y^j(0,y)\| = 1, \end{cases}$$

and by Gronwall's Lemma we deduce that

$$e^{-K_2 t} \leq \|\nabla Y^j(t,y)\| \leq e^{K_2 t},$$

that is $Y^j(t)$ is a bi-Lipschitz homeomorphism with bi-Lipschitz norm controlled by $e^{K_2 t}$. Inserting this information into (5.6) we deduce that, provided T is small enough so that

$$e^{K_2 T} \leq 2, \tag{5.7}$$

it holds

$$\|\rho_t^j\|_{C^{0,\alpha}(\mathbb{T}^2)} \leq e^{K_2 t}\|\rho_0\|_{C^{0,\alpha}(\mathbb{T}^2)} \leq K_0 \qquad \forall\, t \in \left[0, \tfrac{T}{j}\right]. \tag{5.8}$$

(Recall that, by definition, $K_0 = 2\|\rho_0\|_{C^{0,\alpha}(\mathbb{T}^2)}$.)

We now repeat this procedure over the time interval $t \in \left[\frac{T}{j}, 2\frac{T}{j}\right]$. More precisely, for $t \in \left[\frac{T}{j}, 2\frac{T}{j}\right]$ we consider $P_t^{*,j}$ the unique map whose gradient sends $\rho_{T/j}^j$ to dx, we define U_t^j for $t \in \left[\frac{T}{j}, 2\frac{T}{j}\right]$ as in (5.3), we consider its flow $Y^j(t)$, and we use this flow to let $\rho_{T/j}^j$ evolve over the time interval $\left[\frac{T}{j}, 2\frac{T}{j}\right]$. Notice that, thanks to (5.8), $\|\rho_t^j\|_{C^{0,\alpha}(\mathbb{T}^2)} \leq K_0$ so we still have $\|\nabla U_t^j\|_{L^\infty(\mathbb{T}^2)} \leq K_2$. Hence, by the same argument as above,

$$\|\rho_t^j\|_{C^{0,\alpha}(\mathbb{T}^2)} \leq \|\rho_{T/j}^j\|_{C^{0,\alpha}(\mathbb{T}^2)} e^{K_2\left(t-\frac{T}{j}\right)} \qquad \forall\, t \in \left[\tfrac{T}{j}, 2\tfrac{T}{j}\right].$$

In particular, combining this bound with (5.8) and (5.7), we get

$$\|\rho_t^j\|_{C^{0,\alpha}(\mathbb{T}^2)} \leq e^{K_2\left(t-\frac{T}{j}\right)} e^{K_2\frac{T}{j}} \|\rho_0\|_{C^{0,\alpha}(\mathbb{T}^2)} = e^{K_2 t}\|\rho_0\|_{C^{0,\alpha}(\mathbb{T}^2)} \leq K_0 \qquad \forall\, t \in [0, 2\tfrac{T}{j}].$$

Iterating this procedure j times we construct a family $(\rho_t^j, P_t^{*,j})$, with

$$\lambda \leq \rho_t^j \leq \Lambda, \quad \|\rho_t^j\|_{C^{0,\alpha}(\mathbb{T}^2)} \leq K_0, \quad \|D^2 P_t^{*,j}\|_{C^{0,\alpha}(\mathbb{T}^2)} \leq K_1 \qquad \forall\, t \in [0, T], \tag{5.9}$$

such that

$$\begin{cases} \partial_t \rho_t^j + \operatorname{div}(\boldsymbol{U}_t^j \rho_t^j) = 0 & \text{in } [0, T] \times \mathbb{R}^2, \\ \boldsymbol{U}_t^j(y) = \left(\nabla P_t^{*,j}(y) - y\right)^\perp & \text{in } [0, T] \times \mathbb{R}^2, \\ \rho_{iT/j}^j = (\nabla P_t^j)_\sharp dx & \text{for } t \in \left[i\frac{T}{j}, (i+1)\frac{T}{j}\right), \\ \rho_0^j = \rho_0 & \text{on } \mathbb{R}^2. \end{cases} \tag{5.10}$$

Thanks to the bounds (5.9) it is easy to show that, up to subsequences, $(\rho_t^j, P_t^{*,j})$ converge to a solution of (1.6) that will satisfy (5.2) (compare with Sect. 2.4.2). This concludes the proof of the existence part.

5.2 Uniqueness of Smooth Solutions

Let $(\rho_t^1, P_t^{*,1})$ and $(\rho_t^2, P_t^{*,2})$ be two solutions of (1.6) satisfying (5.2). Our goal is to show that they coincide. Because the argument is pretty involved, we shall split it into three steps.

5.2.1 A Gronwall Argument

Recalling that ρ_t^i are given by $Y^i(t)_\# \rho_0$ where $Y^i(t)$ is the flow of $\boldsymbol{U}_t^i = (\nabla P_t^{*,i} - y)^\perp$, $i = 1, 2$ (see Sect. 2.1), it is enough to show that $Y^1(t) = Y^2(t)$. So we compute

$$\begin{aligned} \frac{d}{dt}\int_{\mathbb{T}^2} |Y^1(t) - Y^2(t)|^2\, dy &= 2\int_{\mathbb{T}^2} \left(Y^1(t) - Y^2(t)\right)\cdot\left(\dot Y^1(t) - \dot Y^2(t)\right) dy \\ &= 2\int_{\mathbb{T}^2} \left(Y^1(t) - Y^2(t)\right)\cdot\left(\boldsymbol{U}_t^1(Y^1(t)) - \boldsymbol{U}_t^2(Y^2(t))\right) dy \end{aligned}$$

$$
\begin{aligned}
&= 2\int_{\mathbb{T}^2} \big(Y^1(t) - Y^2(t)\big) \cdot \big(U_t^1(Y^1(t)) - U_t^1(Y^2(t))\big)\, dy \\
&\quad + 2\int_{\mathbb{T}^2} \big(Y^1(t) - Y^2(t)\big) \cdot \big(U_t^1(Y^2(t)) - U_t^2(Y^2(t))\big)\, dy \\
&\le 2\|\nabla U_t^1\|_{L^\infty(\mathbb{T}^2)} \int_{\mathbb{T}^2} \big|Y^1(t) - Y^2(t)\big|^2\, dy \\
&\quad + \int_{\mathbb{T}^2} \big|Y^1(t) - Y^2(t)\big|^2\, dy \\
&\quad + \int_{\mathbb{T}^2} \big|U_t^1(Y^2(t)) - U_t^2(Y^2(t))\big|^2\, dy,
\end{aligned}
$$

where at the last step we used that $2a \cdot b \le |a|^2 + |b|^2$. Notice that (5.2) implies that ∇U_t^1 is bounded, hence the above estimate gives

$$
\begin{aligned}
\frac{d}{dt}\int_{\mathbb{T}^2} \big|Y^1(t) - Y^2(t)\big|^2\, dy &\le C\int_{\mathbb{T}^2} \big|Y^1(t) - Y^2(t)\big|^2\, dy \\
&\quad + \int_{\mathbb{T}^2} \big|U_t^1(Y^2(t)) - U_t^2(Y^2(t))\big|^2\, dy.
\end{aligned}
\tag{5.11}
$$

We now want to bound the last term in the right hand side. For this we first notice that

$$
|U_t^1 - U_t^2| = |(\nabla P_t^{*,1} - y)^\perp - (\nabla P_t^{*,2} - y)^\perp| = |(\nabla P_t^{*,1} - \nabla P_t^{*,2})^\perp| = |\nabla P_t^{*,1} - \nabla P_t^{*,2}|,
\tag{5.12}
$$

hence, recalling that $\rho_t^2 = Y^2(t)_\# \rho_0$, we get

$$
\begin{aligned}
\int_{\mathbb{T}^2} \big|U_t^1(Y^2(t)) - U_t^2(Y^2(t))\big|^2\, dy &\overset{(5.2)}{\le} \frac{1}{\lambda}\int_{\mathbb{T}^2} \big|U_t^1(Y^2(t)) - U_t^2(Y^2(t))\big|^2 \rho_t^2\, dy \\
&\overset{(1.3)}{=} \frac{1}{\lambda}\int_{\mathbb{T}^2} \big|U_t^1 - U_t^2\big|^2 \rho_0\, dy \\
&\overset{(5.2)}{\le} \frac{\Lambda}{\lambda}\int_{\mathbb{T}^2} \big|U_t^1 - U_t^2\big|^2\, dy \\
&\overset{(5.12)}{\le} \frac{\Lambda}{\lambda}\int_{\mathbb{T}^2} \big|\nabla P_t^{*,1} - \nabla P_t^{*,2}\big|^2\, dy.
\end{aligned}
\tag{5.13}
$$

Thus we are left with estimating the L^2 norm of $\nabla P_t^{*,1} - \nabla P_t^{*,2}$.

5.2.2 An Interpolation Argument

To estimate $\|\nabla P_t^{*,1} - \nabla P_t^{*,2}\|_{L^2(\mathbb{T}^2)}$, the idea is to find a curve $[1,2] \ni \theta \mapsto \nabla P_t^{*,\theta}$ which interpolates between these two functions, write

$$\nabla P_t^{*,1} - \nabla P_t^{*,2} = \int_1^2 \partial_\theta \nabla P_t^{*,\theta}\, d\theta$$

so that by Holder's inequality

$$\|\nabla P_t^{*,1} - \nabla P_t^{*,2}\|^2_{L^2(\mathbb{T}^2)} \leq \left(\int_1^2 \|\partial_\theta \nabla P_t^{*,\theta}\|_{L^2(\mathbb{T}^2)}\, d\theta\right)^2 \leq \int_1^2 \|\partial_\theta \nabla P_t^{*,\theta}\|^2_{L^2(\mathbb{T}^2)}\, d\theta, \tag{5.14}$$

and try to control $\|\partial_\theta \nabla P_t^{*,\theta}\|_{L^2(\mathbb{T}^2)}$ with $\|Y^1(t) - Y^2(t)\|_{L^2(\mathbb{T}^2)}$ in order to close the Gronwall argument in (5.11).

To this aim, we consider a curve of measure $[1,2] \ni \theta \mapsto \rho_t^\theta$ (to be chosen) which interpolates between ρ_t^1 and ρ_t^2 and define $\nabla P_t^{*,\theta}$ as the optimal map sending ρ_t^θ onto dx (see Theorem 2.2). Assume that the measures ρ_t^θ satisfy

$$\frac{1}{K_2} \leq \rho_t^\theta \leq K_2, \qquad \|\rho_t^\theta\|_{C^{0,\alpha}(\mathbb{T}^2)} \leq K_2, \tag{5.15}$$

for some universal constant $K_2 > 0$, so that[13]

$$\|D^2 P_t^{*,\theta}\|_{L^\infty(\mathbb{T}^2)} \leq K_3, \qquad \|(D^2 P_t^{*,\theta})^{-1}\|_{L^\infty(\mathbb{T}^2)} \leq K_3. \tag{5.16}$$

Also, assume that there is a vector field $\mathbf{V}_t^\theta$ such that

$$\partial_\theta \rho_t^\theta + \operatorname{div}(\mathbf{V}_t^\theta \rho_t^\theta) = 0 \qquad \text{on } [1,2] \times \mathbb{R}^2. \tag{5.17}$$

(Notice that here t is just a fixed parameter, while θ is playing the role of the time variable.)

Then, by the very same computations as in the proof of Theorem 4.5 we obtain

$$\int_{\mathbb{T}^2} \rho_t^\theta |(D^2 P_t^{*\theta})^{-1/2} \partial_\theta \nabla P_t^{*,\theta}|^2\, dx \leq \int_{\mathbb{T}^2} \rho_t^\theta |(D^2 P_t^{*,\theta})^{1/2} \mathbf{V}_t^\theta|^2\, dx$$

(compare with (4.36)), and using (5.15) and (5.16) we deduce that

$$\int_{\mathbb{T}^2} |\partial_\theta \nabla P_t^{*,\theta}|^2\, dx \leq K_4 \int_{\mathbb{T}^2} |\mathbf{V}_t^\theta|^2 \rho_t^\theta\, dx,$$

[13]The bound on $D^2 P_t^{*,\theta}$ follows by the $C^{2,\alpha}$ regularity for Monge-Ampère [10], while the bound for $(D^2 P_t^{*,\theta})^{-1}$ follows exactly as in the proof of (4.21).

that combined with (5.13) and (5.14) gives

$$\int_{\mathbb{T}^2} \left|U_t^1(Y^2(t)) - U_t^2(Y^2(t))\right|^2 dy \le K_4 \frac{\Lambda}{\lambda} \int_1^2 \left(\int_{\mathbb{T}^2} |V_t^\theta|^2 \rho_t^\theta \, dy \right) d\theta. \tag{5.18}$$

Hence, our goal is to choose $(\rho_t^\theta, V_t^\theta)$ in such a way that (5.15)–(5.17) hold, and the right hand side above is controlled by $\| Y^1(t) - Y^2(t)\|_{L^2(\mathbb{T}^2)}$.[14]

5.2.3 Construction of the Interpolating Curve

The key observation is that, since $Y^1(t)_\# \rho_0 = \rho_t^1$ and $Y^2(t)_\# \rho_0 = \rho_t^2$, the map $S_t := Y^2(t) \circ [Y^1(t)]^{-1}$ satisfies

$$(S_t)_\# \rho_1^t = \rho_2^t.$$

Hence, if $T_t = \nabla \Phi_t : \mathbb{T}^2 \to \mathbb{T}^2$ denotes the optimal transport map from ρ_t^1 to ρ_t^2, by the definition of optimal transport (see Sect. 2.2) we have

$$\int_{\mathbb{T}^2} |S_t - y|^2 \rho_t^1(y) \, dy \ge \int_{\mathbb{T}^2} |T_t - y|^2 \rho_t^1(y) \, dy.$$

Also, since $[Y^1(t)^{-1}]_\# \rho_t^1 = \rho_0$,

$$\begin{aligned}
\int_{\mathbb{T}^2} |S_t - y|^2 \rho_t^1(y) \, dy &= \int_{\mathbb{T}^2} \left| Y^2(t) \circ [Y^1(t)]^{-1} - y\right|^2 \rho_t^1(y) \, dy \\
&\overset{(1.3)}{=} \int_{\mathbb{T}^2} | Y^2(t) - Y^1(t)|^2 \rho_0(y) \, dy \\
&\overset{(5.1)}{\le} \Lambda \int_{\mathbb{T}^2} | Y^2(t) - Y^1(t)|^2 \, dy,
\end{aligned}$$

therefore

$$\Lambda \int_{\mathbb{T}^2} | Y^2(t) - Y^1(t)|^2 \, dy \ge \int_{\mathbb{T}^2} |T_t - y|^2 \rho_t^1(y) \, dy. \tag{5.19}$$

---

[14] The reader familiar with optimal transport theory may recognize in (5.18) the dynamic formulation of optimal transportation discovered by Benamou and Brenier [6]:

$$\min\left\{ \int_1^2 \left(\int_{\mathbb{T}^2} |V_t^\theta|^2 \rho_t^\theta \, dx \right) d\theta \, : \, (\rho_t^\theta, V_t^\theta) \text{ satisfy } (5.17) \right\} = \min\left\{ \int_{\mathbb{R}^n} |S(x) - x|^2 \, d\mu(x) \, : \, S_\# \rho_t^1 = \rho_t^2 \right\}.$$

Although we shall not use this fact, the argument in Sect. 5.2.3 is strongly inspired by it.

Also, since both ρ_t^1 and ρ_t^2 satisfy (5.2), the bounds

$$\|D^2\Phi_t\|_{C^{0,\alpha}(\mathbb{T}^2)} \le \hat{K}, \qquad \|(D^2\Phi_t)^{-1}\|_{L^\infty(\mathbb{T}^2)} \le \hat{K} \tag{5.20}$$

hold (compare with (5.16), see also Footnote 13).

We now would like to relate V_t^θ to $T_t(y) - y$, and this suggests the following definition of ρ_t^θ (as already mentioned in Footnote 14, this is strongly inspired by Benamou and Brenier [6]):

$$\rho_t^\theta := \big[y + (\theta - 1)(T_t(y) - y)\big]_{\#}\rho_1^t \qquad \forall \theta \in [1,2],$$

or equivalently, since $T_t = \nabla\Phi_t$,

$$\rho_t^\theta = [\nabla\Phi_t^\theta]_{\#}\rho_1^t, \qquad \Phi_t^\theta := (2-\theta)\frac{|y|^2}{2} + (\theta - 1)\Phi_t.$$

Let

$$\Phi_t^{\theta,*}(y) := \sup_{x \in \mathbb{R}^2}\{x \cdot y - \Phi_t^\theta(x)\}.$$

Recalling that $\nabla\Phi_t^{\theta,*} = (\nabla\Phi_t^\theta)^{-1}$ (see (1.5)), one can check that with these definitions the following properties hold[15]:

$$\begin{cases} \text{(A)} \quad \text{(5.17) holds with } V_t^\theta := (T_t - y) \circ \nabla\Phi_t^{\theta,*}, & \\ \text{(B)} \quad \int_{\mathbb{T}^2} |V_t^\theta|^2 \rho_t^\theta \, dy = \int_{\mathbb{T}^2} |T_t - y|^2 \, \rho_t^1 \, dy & \forall \theta \in [1,2], \\ \text{(C)} \quad \det(D^2\Phi_t^\theta) = \frac{\rho_t^1}{\rho_t^\theta \circ \nabla\Phi_t^\theta} & \forall \theta \in [1,2]. \end{cases} \tag{5.21}$$

5.2.4 Bounds on the Interpolating Curve: Proof of (5.15)

We now prove that the measures ρ_t^θ satisfy all properties in (5.15).

First of all we notice that, thanks to (5.20),

$$\frac{1}{\hat{K}}\,\mathrm{Id} \le D^2\Phi_t \le \hat{K}\,\mathrm{Id},$$

[15]Property (A) follows by a direct computation very similar to what we already did in Sect. 2.1 to show that $Y(t)_{\#}\bar{\sigma}$ solves (2.1). Property (B) is a direct consequence of (1.3) and the fact that $[(\nabla\Phi_t^\theta)^{-1}]_{\#}\rho_t^\theta = \rho_1^t$, while (C) follows by (2.7). We leave the details to the interested reader.

therefore, since $D^2\Phi_t^\theta = (2-\theta)\mathrm{Id} + (\theta-1)D^2\Phi_t$, it follows immediately that

$$\frac{1}{\hat{K}}\,\mathrm{Id} \le D^2\Phi_t^\theta \le \hat{K}\,\mathrm{Id} \qquad \forall\theta\in[1,2]. \tag{5.22}$$

In particular $\det(D^2\Phi_t^\theta) \in \left[\frac{1}{\hat{K}^n}, \hat{K}^n\right]$, that combined with (5.21)-(C) and the fact that ρ_t^1 satisfies (5.2) gives

$$\rho_t^\theta = \frac{\rho_t^1}{\det(D^2\Phi_t^\theta)} \circ \nabla\Phi_t^{\theta,*} \in \left[\frac{\lambda}{\hat{K}^n}, \frac{\hat{K}^n}{\lambda}\right]. \tag{5.23}$$

Also, the Hölder continuity of $D^2\Phi_t$ (see (5.20)) implies that $D^2\Phi_t^\theta \in C^{0,\alpha}$, from which it follows that

$$\left\|\det(D^2\Phi_t^\theta)\right\|_{C^{0,\alpha}(\mathbb{T}^2)} \le \hat{K}_0,$$

so by (5.2) and the fact that $\det(D^2\Phi_t^\theta) \ge 1/\hat{K}^n$ we get

$$\left\|\frac{\rho_t^1}{\det(D^2\Phi_t^\theta)}\right\|_{C^{0,\alpha}(\mathbb{T}^2)} \le \hat{K}_1.$$

Finally, it suffices to observe that $\|D^2\Phi_t^{\theta,*}\| \le \hat{K}$ (this simply follows from (5.22) and (3.5)) to deduce that $\nabla\Phi_t^{\theta,*}$ is uniformly Lipschitz, thus

$$\left\|\frac{\rho_t^1}{\det(D^2\Phi_t^\theta)} \circ \nabla\Phi_t^{\theta,*}\right\|_{C^{0,\alpha}(\mathbb{T}^2)} \le \hat{K}_2.$$

Recalling (5.23), this concludes the proof of (5.15).

5.2.5 Conclusion

The fact that the measures ρ_t^θ satisfy the properties in (5.15) allows us to justify all the previous computations. In particular, thanks to (5.19), (5.18), and (5.21)-(B), we get

$$\int_{\mathbb{T}^2} \left|U_t^1(Y^2(t)) - U_t^2(Y^2(t))\right|^2 dy \le K_4\,\frac{\Lambda^2}{\lambda} \int_{\mathbb{T}^2} |Y^2(t) - Y^1(t)|^2\,dy.$$

Inserting this bound into (5.11), we finally obtain

$$\frac{d}{dt}\int_{\mathbb{T}^2} \left|Y^1(t) - Y^2(t)\right|^2 dy \le \bar{C} \int_{\mathbb{T}^2} |Y^2(t) - Y^1(t)|^2\,dy,$$

so by Gronwall's inequality

$$\int_{\mathbb{T}^2} |Y^1(t) - Y^2(t)|^2 \, dy \le e^{\bar{C}t} \int_{\mathbb{T}^2} |Y^1(0) - Y^2(0)|^2 \, dy = 0,$$

as desired.

6 Open Problems

In this last section we state some open problems related to the Monge-Ampère and semigeostrophic equations.

1. Our global existence result for weak solutions of SG was based on regularity results for Monge-Ampère that are valid in every dimension. However, the regularity theory for Monge-Ampère provides stronger results in 2-D. For instance, Alexandrov showed in [4] (see also [25, Theorem 2.1]) that a convex function $\phi : \mathbb{R}^2 \to \mathbb{R}$ is continuous differentiable provided the upper bound $\det(D^2\phi) \le \Lambda$ holds (this result is false when $n \ge 3$, see [35]). Hence, in relation to the theorem proved in Sect. 4.1, a natural question becomes the following:

   Is it possible to prove $W^{2,1}_{\mathrm{loc}}$ *regularity of* ϕ *in the* 2-*D case assuming only an upper bound on* $\det(D^2\phi)$?

   Apart from its own interest, such a result could help in extending Theorem 3.1 outside of the periodic setting.
2. As shown in Sect. 5, the existence of smooth solutions for the dual SG system is known only for short time. However, for the 2-D incompressible Euler equations it is well-known that smooth solutions exist globally in time (see for instance [8, Corollary 3.3]). By the analogy between the dual SG system and the Euler equations (see Sect. 2.3) one may hope to say that global smooth solutions exist also for the dual SG system, at least for initial data which are sufficiently close to 1 in some strong norm. Whether this fact holds true is an interesting open problem.
3. As proved in [3], the results described here can be extended to the case when the domain is the whole $\mathbb{R}^2$,[16] provided the initial datum $\rho_0 = (\nabla P_0)_\sharp dx\lfloor_\Omega$ is strictly positive on the whole space. It would nice to remove this assumption in order to deal with the case when ρ_0 is compactly supported (which is the most interesting case from a physical point of view). However, the nontrivial evolution of the support of the solution ρ_t does not permit to apply the regularity results from [17, 20, 31], so completely new ideas need to be introduced in order to prove existence of distributional solutions to (1.1) in this case. As already mentioned above, solving Problem 1 could be extremely helpful in this direction.

[16] To be precise, the results in [3] are three dimensional, but they also hold in 2-D.

Acknowledgements Alessio Figalli is supported by NSF Grant DMS-1262411.

References

1. L. Ambrosio, Transport equation and Cauchy problem for *BV* vector fields. Invent. Math. **158**, 227–260 (2004)
2. L. Ambrosio, M. Colombo, G. De Philippis, A. Figalli, Existence of Eulerian solutions to the semigeostrophic equations in physical space: the 2-dimensional periodic case. Commun. Partial Differ. Equ. **37**(12), 2209–2227 (2012)
3. L. Ambrosio, M. Colombo, G. De Philippis, A. Figalli, A global existence result for the semigeostrophic equations in three dimensional convex domains. Discrete Contin. Dyn. Syst. **34**(4), 1251–1268 (2014)
4. A.D. Alexandrov, Smoothness of the convex surface of bounded Gaussian curvature. C. R. (Doklady) Acad. Sci. URSS (N. S.) **36**, 195–199 (1942)
5. J.-D. Benamou, Y. Brenier, Weak existence for the semigeostrophic equation formulated as a coupled Monge-Ampère/transport problem. SIAM J. Appl. Math. **58**, 1450–1461 (1998)
6. J.-D. Benamou, Y. Brenier, A computational fluid mechanics solution to the Monge-Kantorovich mass transfer problem. Numer. Math. **84**(3), 375–393 (2000)
7. J. Bergh, J. Löfström, *Interpolation Spaces. An Introduction*. Grundlehren der Mathematischen Wissenschaften, vol. 223 (Springer, Berlin, 1976), x+207 pp.
8. A.L. Bertozzi, A.J. Majda, *Vorticity and Incompressible Flow*. Cambridge Texts in Applied Mathematics, vol. 27 (Cambridge University Press, Cambridge, 2002)
9. Y. Brenier, Polar factorization and monotone rearrangement of vector-valued functions. Commun. Pure Appl. Math. **44**(4), 375–417 (1991)
10. L. Caffarelli, Interior $W^{2,p}$ estimates for solutions of the Monge-Ampère equation. Ann. Math. (2) **131**(1), 135–150 (1990)
11. L. Caffarelli, Some regularity properties of solutions to Monge-Ampère equations. Commun. Pure Appl. Math. **44**, 965–969 (1991)
12. L. Caffarelli, C. Gutierrez, Real analysis related to the Monge-Ampère equation. Trans. Am. Math. Soc. **348**(3), 1075–1092 (1996)
13. D. Cordero Erausquin, Sur le transport de mesures périodiques. C. R. Acad. Sci. Paris Sér. I Math. **329**, 199–202 (1999)
14. M. Cullen, *A Mathematical Theory of Large-Scale Atmosphere/Ocean Flow* (Imperial College Press, London, 2006)
15. M. Cullen, M. Feldman, Lagrangian solutions of semigeostrophic equations in physical space. SIAM J. Math. Anal. **37**, 1371–1395 (2006)
16. S. Daneri, A. Figalli, Variational models for the incompressible Euler equations, in *HCDTE Lecture Notes. Part II. Nonlinear Hyperbolic PDEs, Dispersive and Transport Equations*. AIMS Ser. Appl. Math., vol. 7 (Am. Inst. Math. Sci. (AIMS), Springfield, MO, 2013), 51 pp
17. G. De Philippis, A. Figalli, $W^{2,1}$ regularity for solutions of the Monge-Ampère equation. Invent. Math. **192**(1), 55–69 (2013)
18. G. De Philippis, A. Figalli, Second order stability for the Monge-Ampère equation and strong Sobolev convergence of optimal transport maps. Anal. Partial Differ. Equ. **6**(4), 993–1000 (2013)
19. G. De Philippis, A. Figalli, The Monge-Ampère equation and its link to optimal transportation. Bull. Am. Math. Soc. (N.S.), **51**(4), 527–580 (2014)
20. G. De Philippis, A. Figalli, O. Savin, A note on interior $W^{2,1+\epsilon}$ estimates for the Monge-Ampère equation. Math. Ann. **357**(1), 11–22 (2013)
21. L.C. Evans, *Partial Differential Equations*. Graduate Studies in Mathematics, vol. 19, 2nd edn. (American Mathematical Society, Providence, 2010), xxii+749 pp
22. M. Feldman, A. Tudorascu, On Lagrangian solutions for the semi-geostrophic system with singular initial data. SIAM J. Math. Anal. **45**(3), 1616–1640 (2013)

23. M. Feldman, A. Tudorascu, On the Semi-Geostrophic system in physical space with general initial data (2014, Preprint)
24. A. Figalli, Sobolev regularity for the Monge-Ampère equation, with application to the semi-geostrophic equations. Zap. Nauchn. Sem. S.-Peterburg. Otdel. Mat. Inst. Steklov. (POMI), **411** (2013), Teoriya Predstavlenii, Dinamicheskie Sistemy, Kombinatornye Metody. XXII, 103–118, 242; translation in J. Math. Sci. (N. Y.), **196**(2), 175–183 (2014)
25. A. Figalli, G. Loeper, C^1 regularity of solutions of the Monge-Ampère equation for optimal transport in dimension two. Calc. Var. Partial Differ. Equ. **35**(4), 537–550 (2009)
26. C. Gutierrez, *The Monge-Ampére Equation*. Progress in Nonlinear Differential Equations and Their Applications, vol. 44 (Birkhäuser Boston, Inc., Boston, 2001)
27. C. Gutierrez, Q. Huang, Geometric properties of the sections of solutions to the Monge-Ampère equation. Trans. Am. Math. Soc. **352**(9), 4381–4396 (2000)
28. F. John, *Extremum Problems with Inequalities as Subsidiary Conditions*, in Studies and Essays Presented to R. Courant on his 60th Birthday, January 8, 1948 (Interscience, New York, 1948), pp. 187–204
29. G. Loeper, On the regularity of the polar factorization for time dependent maps. Calc. Var. Partial Differ. Equ. **22**, 343–374 (2005)
30. G. Loeper, A fully non-linear version of the incompressible Euler equations: the semi-geostrophic system. SIAM J. Math. Anal. **38**, 795–823 (2006)
31. T. Schmidt, $W^{2,1+\epsilon}$ estimates for the Monge-Ampère equation (2012, Preprint)
32. J. Simon, Compact sets in the space $L^p(0, T; B)$. Ann. Mat. Pura Appl. (4) **146**, 65–96 (1987)
33. E. Stein, *Harmonic Analysis: Real-Variable Methods, Orthogonality, and Oscillatory Integrals. With the Assistance of Timothy S. Murphy*. Princeton Mathematical Series, 43. Monographs in Harmonic Analysis, III (Princeton University Press, Princeton, 1993)
34. C. Villani, *Optimal Transport. Old and New*. Grundlehren der Mathematischen Wissenschaften [Fundamental Principles of Mathematical Sciences], vol. 338 (Springer, Berlin, 2009)
35. X.-J. Wang, Some counterexamples to the regularity of Monge-Ampère equations. Proc. Am. Math. Soc. **123**(3), 841–845 (1995)

On Some Elliptic and Parabolic Equations Related to Growth Models

Ireneo Peral

DEDICATORY
To Iria, Laura and Oscar, the owners of my heart

Presentation

We present in this note the contents of the course delivered in the *C.I.M.E.* program from 2nd to 6th of June, 2014, at Cetraro, Italy. The course consisted of a series of four lessons of hour and a half each one, about the study of some mathematical problems that appears as model of growth phenomena. There exists a large amount of problems coming from the Physics and describing phenomena of *growth*. If one gives a look to the literature on this subject there are a lot of different proposals according to the special characteristics of the situation to describe. The physical models are *stochastic* across the source terms, but here we will only describe some results for *deterministic* models.

Mainly we will consider the following kind of models.

1. The classical Kardar-Pasisi-Zhang model, analyzing multiplicity of solutions. See [89] to see the ideas of the modelization.
2. The Kardar-Pasisi-Zhang model in porous media. See [21] for the modelization of some particular cases.
3. A variational model that tries to describe *epitaxial growth* of crystals. See [65] for some ideas concerning this model.

CIME Cetraro 2014.

I. Peral (✉)
Departamento de Matemáticas, Universidad Autónoma de Madrid, Madrid, Spain
e-mail: ireneo.peral@uam.es

A. Farina, E. Valdinoci (eds.), *Partial Differential Equations and Geometric Measure Theory*, Lecture Notes in Mathematics 2211,
https://doi.org/10.1007/978-3-319-74042-3_2

The more accurate to the physical problems seems to be the *stochastic* setting in which, as far we know, there is still a huge amount of work to do and it is far away from a satisfactory situation. As an example of outstanding achievement in this direction and for the so called KPZ model, we would like to mention the recent paper by Martin Hairer, [83], which is a milestone by creating an appropriate calculus to handle *nonlinear stochastic* problems.

Even in the deterministic setting the mathematical methods to be used are of a great diversity. For this reason these notes can be useful to the reader as a good training in different subjects of mathematics which are interesting themselves.

Respect to the problems in the part (3) above can be seen, for instance, the papers [73, 74, 92, 93] and [147] with a different type of models from the one studied here.

The results presented in these Lecture Notes are based on papers of the author carried out in the last years in collaboration with my colleagues: B. Abdellaoui, A. Dall'Aglio, D. Giachetti, C. Escudero, F. Ferrari, F. Gazzola, M. Medina, S. Segura and M. Walias (see [4, 5, 7–9, 59, 66, 72] and [67]). I would like to express my warm gratitude to all of them.

I would also thank to Alberto Farina and Enrico Valdinoci for their kind invitation to participate in the C.I.M.E. course.

The organization of this Lecture Notes is in four chapters, each one corresponds to a Lecture of the course.

1 Kardar-Parisi-Zhang Model of Growing Interfaces: Stationary Model

1.1 Introduction

1.1.1 Kardar-Parisi-Zhang Model of Growing Interfaces

We consider the following problem

$$\begin{cases} u_t - \Delta u = |\nabla u|^2 + f(x,t) & \text{in } Q \equiv \Omega \times (0,+\infty) \\ u(x,t) = 0 & \text{on } \partial\Omega \times (0,+\infty), \\ u(x,0) = u_0(x) & \text{in } \Omega, \end{cases} \tag{1.1}$$

where Ω is a bounded regular domain and f, and u_0 are positive functions satisfying some hypotheses that we will specify later. This parabolic equation appears in the physical theory of growth and roughening of surfaces, where it is known as the Kardar-Parisi-Zhang equation, see [89]. The nonlinear term appears by approximating $\sqrt{1+|\nabla u|^2} \approx 1 + \frac{1}{2}|\nabla u|^2$. That is, only small gradients are considered.

In the form

$$u_t - \epsilon^2 \Delta u = |\nabla u|^2,$$

it may be viewed as the viscosity approximation as $\varepsilon \to 0^+$ of Hamilton-Jacobi type equations from stochastic control theory (see [104]). A modification of the problem is also studied by Berestycki, Kamin, and Sivashinsky as a model in flame propagation (see [28]).

Notice that this model is a crossroad of several problems. For instance, with some hypotheses on u, if we perform the Hopf-Cole change of variables, $v = e^u - 1$ we get the equation

$$v_t - \Delta v = f(v+1).$$

In some textbooks this equation is used to study growth of brain tumors (see [116, 117]).

If we derive the equation (E) with respect to the spacial variables and calling $v_i = u_{x_i}$ we get

$$v_{i\,t} - \Delta v_i = 2 v_i \sum_1^N v_{i\,x_j}, \qquad \mathbf{v} = \nabla u,$$

which in dimension $N = 1$ is just the Burgers's equation. If $N > 1$ is the Euler equation without the pressure that we can recover putting a source term in equation (E).

In this chapter we will study some properties of the associated stationary problem while the evolution problem will be studied in Sect. 2.

This chapter is mainly a part of the paper [4].

1.1.2 The Stationary Problem: Planing for Sect. 1

This chapter is devoted to explain some results concerning nonlinear elliptic equations of the form

$$\begin{cases} -\Delta u = |\nabla u|^2 + \lambda f(x) & \text{in } \Omega \\ \quad u = 0 & \text{on } \partial\Omega, \end{cases} \tag{1.2}$$

where Ω is a bounded open set in $\mathbb{R}^N$, λ is a positive constant and $f(x)$ is a positive measurable function. We will assume that Ω has a smooth enough boundary, as an example, the interior sphere condition is sufficient to do all the arguments below. Equations of the form (1.2) have been widely studied in the literature. Existence results for problem (1.2) start from the classic references [102] and [96]. Later on, many authors have been considering elliptic equations with first order terms having quadratic growth with respect to the gradients (see for instance [35–38, 40, 43, 52, 55, 69, 70, 82, 89, 90, 115, 129, 130, 135, 143] and references therein). It is well known that in this case (see [90] and [69]) the change of variable $v = e^u - 1$ leads to the linear equation

$$\begin{cases} -\Delta v = \lambda f(x)(v+1) & \text{in } \Omega \\ \quad v = 0 & \text{on } \partial\Omega, \end{cases} \tag{1.3}$$

which admits a unique solution in $W_0^{1,2}(\Omega)$ provided $f \in L^{N/2}$ and λ is small enough. It is also known that the smallness condition on λ is necessary in order to have existence. This means that, for every $f(x) \geq 0$, with $f \not\equiv 0$, there is no solution of problem (1.3) for λ large. Therefore Eq. (1.2) has no solution in the space of functions u such that $e^u - 1 \in W_0^{1,2}(\Omega)$ (see also [69] for a detailed result in this direction). A first contribution of our paper is a non-existence result in the larger space $W_0^{1,2}(\Omega)$ when λ is large. More precisely, if $f(x)$ is a locally integrable function, verifying

(A) There exists $\phi_0 \in C_0^\infty(\Omega)$ such that $\displaystyle\int_\Omega |\nabla\phi_0|^2 dx < \lambda \int_\Omega f\phi_0^2 dx < +\infty$;

then we show that problem (1.2) admits no solution in $W_0^{1,2}(\Omega)$ for such λ. We will also analyze the existence and nonexistence under regularity condition on f.

In [70] is showed an example of no uniqueness of solutions of (1.2).

For instance, if $N > 2$, the functions

$$u_m(x) = \log\left(\frac{|x|^{2-N} - m}{1-m}\right) \in W_0^{1,2}(B_1), \;\; 0 \leq m < 1,$$

all solve the equation $-\Delta u = |\nabla u|^2$ in the unit ball $B_1 = \{x \in \mathbb{R}^N \;:\; |x| < 1\}$ (though only the zero function satisfies $e^u - 1 \in W_0^{1,2}(B_1)$).

One of the main aims of this chapter is to characterize this non-uniqueness phenomenon, and to show that *every* solution of problem (1.2) comes from a solution of a linear problem with measure data, after a suitable change of variable. The first step is to show that *all solutions* of Eq. (1.2), also satisfy some exponential integrability (independently on the regularity of $f(x)$, provided this function is nonnegative). More precisely they verify

$$e^{\delta u} - 1 \in W_0^{1,2}(B_1), \quad \text{for every } \delta < \frac{1}{2}.$$

Note that the bound on δ is sharp by the previous example. This subtle lack of summability of the exponential of a solution will be the key of the *wild nonuniqueness* result that will be obtained below.

The proof of this regularity result shows that in fact the *regularizing* term is the right-hand side of the equation, rather than, as usually happens, the diffusion term. Using this regularity result we show that if we perform by a convenient approximation argument the change of variable $v = e^u - 1$, then the new function v still lives in a larger Sobolev space, that is, $v \in W_0^{1,q}(\Omega)$, for every $q < \frac{N}{N-1}$. This regularity result is deeply related with the Stampacchia results to obtain solutions of elliptic equation with measure data. In fact we are able to prove that v is a distributional solution of the problem

$$\begin{cases} -\Delta v = \lambda f(x)(v+1) + \mu_s & \text{in } \Omega \\ v = 0 & \text{on } \partial\Omega, \end{cases} \tag{1.4}$$

where μ_s is a bounded positive Radon measure which is concentrated on a set of capacity zero.

On the other hand, we can also prove a result in the opposite direction, that is, if f is a nonnegative function such that $f \in L^q(\Omega)$, with $q > \frac{N}{2}$, and if λ is small enough, and if μ_s is a bounded positive Radon measure, then problem (1.4) has a unique solution, a result which was proved for bounded f by Radulescu-Willem [134] (see also Orsina [120]). Then, if μ_s is concentrated on a set of zero capacity, we will show that $u = \log(1 + v) \in W_0^{1,2}(\Omega)$ is a solution of problem (1.2).

Therefore, summarizing, solutions to problem (1.2) are in one-to-one correspondence with measures concentrates in a subset of Ω with zero newtonian capacity.

Further results will be considered in Sect. 1.4 below.

The non-uniqueness results are based on the following Picone type inequality (see [127] and [1] for some extensions) that we prove for the reader convenience.

Theorem 1.1 *If $u \in W_0^{1,2}(\Omega)$, $u \geq 0$, $v \in W_0^{1,2}(\Omega)$, $-\Delta v \geq 0$ is a bounded Radon measure, $v|_{\partial\Omega} = 0$, $v \geq 0$ and not identically zero, then*

$$\int_\Omega |\nabla u|^2 \geq \int_\Omega \left(\frac{u^2}{v}\right)(-\Delta v).$$

Proof To prove the result above we need the following pointwise result that in the quadratic case is an easy computation.

Lemma 1.2 *Let $v > 0$, $u \geq 0$ two differentiable functions, let*

$$L(u, v) = |\nabla u|^2 + \frac{u^2}{v^2}|\nabla v|^2 - 2\frac{u}{v}\langle \nabla v, \nabla u\rangle.$$

$$R(u, v) = |\nabla u|^2 - \nabla\left(\frac{u^2}{v}\right)\nabla v.$$

Then

1. $L(u, v) = R(u, v)$

2. *$L(u,v)\geq 0$ and $L(u,v)=0$, a.e. in Ω if and only if $u=kv$ in each connected component of Ω.*

We first prove the following particular case.

Lemma 1.3 *Let $v\in W^{1,2}(\Omega)$ be such that $v\geq\delta>0$ in Ω. Then for all $u\in C_0^\infty(\Omega)$, $u\geq 0$*

$$\int_\Omega |\nabla u|^2 \geq \int_\Omega \left(\frac{u^2}{v}\right)(-\Delta v).$$

Proof Since $v\in W^{1,2}(\Omega)$ and $v\geq\delta>0$ in Ω there exist a family of regular function $\{v_n\}$ such that

$$\begin{cases} v_n \to v \text{ in } W^{1,2}(\Omega), \;\; v_n\in C^1(\Omega),\\ v_n\to v,\; a.e, \text{ and } v_n>\frac{\delta}{2} \text{ in } \Omega. \end{cases} \tag{1.5}$$

As a consequence of the continuity of the $-\Delta$ as an operator from $W^{1,2}(\Omega)$ to $W^{-1,2}$, we obtain that $-\Delta v_n\to -\Delta v$ in $W^{-1,2}$. By Picone Identity applied to v_n we get

$$|\nabla u|^2 \geq \langle \nabla\left(\frac{u^2}{v_n}\right), \nabla v_n\rangle.$$

Since

$$\begin{aligned}\int_\Omega -\Delta v_n\left(\frac{u^2}{v_n}\right) &= \int_\Omega \langle \nabla v_n, \nabla\left(\frac{u^2}{v_n}\right)\rangle\\ &= 2\int_\Omega \frac{u}{v_n}\langle\nabla v_n,\nabla u\rangle - \int_\Omega \frac{u^2}{v_n^2}|\nabla v_n|^2.\end{aligned}$$

Using the hypothesis on v_n and by the Lebesgue *dominated convergence theorem* we obtain

$$\int_\Omega |\nabla u|^2 \geq \int_\Omega \left(\frac{-\Delta v}{v}\right)u^2,\; u\in C_0^\infty(\Omega),\;\; u\geq 0.$$

□

End of the Proof of Theorem 1.1.

By the strong maximum principle $v>0$ in Ω. We call $v_m(x)=v(x)+\frac{1}{m}$, $m\in\mathbb{N}$. In this way we have that $\Delta v_m=\Delta v$ and $\{v_m\}$ converges in $W^{1,2}(\Omega)$ and almost everywhere. Therefore, by using Lemma 1.3, we obtain the result for $\phi\in C_0^\infty(\Omega)$, $\phi\geq 0$. Now the general case follow by a density argument, namely, if $u_n\to u$ in

$W_0^{1,2}$, $u_n \in C_0^\infty(\Omega)$ and $u_n \geq 0$, then we have that

$$\int_\Omega |\nabla u_n|^2 \geq \int_\Omega \Big(\frac{-\Delta v_n}{v_n}\Big)u_n^2 = \int_\Omega (\frac{-\Delta v}{v_n})u_n^2.$$

By hypothesis and Fatou's Lemma we conclude. □

Remarks 1.4 Consider the eigenvalue problem

$$\begin{cases} -\Delta_p\phi_1 = \lambda_1\rho(x)\phi_1, \ \phi_1 \geq 0 \text{ in } \Omega, \\ \phi_1 = 0, \ \text{on } \partial\Omega \end{cases}$$

where $\rho \geq 0$ and $\rho \in L^r(\Omega)$ for some $r \geq N/2$. Take $v := \phi_1$ in Theorem (1.1). Then Picone inequality becomes

$$\lambda_1 \int_\Omega u^2\rho(x)dx \leq \int_\Omega |\nabla u|^2 dx,$$

that is, the Poincaré inequality for the weight ρ.

1.2 *Analysis of the Solutions in* $W_0^{1,2}(\Omega)$

Consider the problem

$$\begin{cases} -\Delta u = |\nabla u|^2 + \lambda f(x) & \text{in } \Omega \\ u = 0 & \text{on } \partial\Omega, \end{cases} \tag{1.6}$$

where $\lambda > 0$ and $f \in L^m(\Omega)$, $m \geq \dfrac{N}{2}$, $f(x) \geq 0$. Note that, in order to be a solution in the sense of distributions, a function u must be in $W_0^{1,2}(\Omega)$.

1.2.1 Existence and Nonexistence

Assume that $e^u - 1 \in W^{1,2}(\Omega)$ and perform the change of variable

$$v = e^u - 1.$$

Then problem (1.6) becomes

$$\begin{cases} -\Delta v = \lambda f(x)(v+1) & \text{in } \Omega \\ v = 0 & \text{on } \partial\Omega, \end{cases} \tag{1.7}$$

where $f \in L^m(\Omega)$, $m \geq \frac{N}{2}$, $f(x) \geq 0$. Such change of variable is usually called in the literature as Hopf-Cole change.

It is well known that this problem admits a unique solution provided λ is small enough. As a straightforward consequence we obtain the following result.

Theorem 1.5 *If λ is small enough, there exists a unique solution to problem* (1.6) *such that $e^u - 1 \in W_0^{1,2}(\Omega)$.*

Next we will study a deeper existence and nonexistence result according with some hypothesis on f and the size of λ. Assume that f is a measurable, non-negative function such that f satisfies the following property:

(A) There exists $\phi_0 \in C_0^\infty(\Omega)$ such that $\displaystyle\int_\Omega |\nabla\phi_0|^2 dx < \lambda \int_\Omega f\phi_0^2 dx < +\infty$;

then we have the following nonexistence result.

Theorem 1.6 *If λ, f verify the hypothesis* (**A**) *above, then problem* (1.6) *has no solution.*

Proof By contradiction, assume that problem (1.6) has a solution u, then multiplying by ϕ_0^2 we obtain that

$$2\int_\Omega \phi_0 \nabla\phi_0 \nabla u\, dx = \int_\Omega \phi_0^2 |\nabla u|^2 dx + \lambda \int_\Omega f\phi_0^2 dx.$$

Hence we conclude that

$$\lambda \int_\Omega f\phi_0^2 dx = 2\int_\Omega \phi_0 \nabla\phi_0 \nabla u dx - \int_\Omega \phi_0^2 |\nabla u|^2 dx \leq \int_\Omega |\nabla\phi_0|^2 dx$$

a contradiction with the definition of ϕ_0. □

According with the previous result is natural to define for $f \in L^1(\Omega)$ a non-negative function the infimum of the following Rayleigh quotient,

$$\lambda_1(f) = \inf_{\phi \in W_0^{1,2}(\Omega)} \frac{\displaystyle\int_\Omega |\nabla\phi|^2 dx}{\displaystyle\int_\Omega f\phi^2 dx} \geq 0. \tag{1.8}$$

We consider the following hypothesis

(B) $\lambda_1(f) > 0.$

Notice that if f satisfies the hypothesis (**B**), $W_0^{1,2}(\Omega)$ is continuously imbedded in $L^2(\Omega, f(x)dx)$, moreover, using the Cauchy-Schwartz inequality for the measure

$f(x)dx$ and hypothesis (**B**) we obtain

$$\int_\Omega vf(x)dx \le \left(\int_\Omega f(x)dx\right)^{\frac{1}{2}}\left(\int_\Omega v^2 f(x)dx\right)^{\frac{1}{2}}$$
$$\le \lambda_1(f)^{-1/2}\left(\int_\Omega f(x)dx\right)^{\frac{1}{2}}\left(\int_\Omega |\nabla v|^2 dx\right)^{\frac{1}{2}}.$$

That is, hypothesis (**B**) implies that $f \in W^{-1,2}(\Omega)$. As a consequence we can formulate the next result.

Theorem 1.7 *Assume that* (**B**) *holds, then problem* (1.6) *has no solution in* $W_0^{1,2}(\Omega)$ *for* $\lambda > \lambda_1(f)$ *and has a unique solution* u *such that* $e^u - 1 \in W_0^{1,2}(\Omega)$ *for* $\lambda < \lambda_1(f)$.

Proof If $\lambda > \lambda_1(f)$, then by a density argument we can show that condition (**A**) holds. Therefore, by Theorem 1.6 we obtain that problem (1.6) has no solution. We prove now the existence result. Assume that $\lambda < \lambda_1(f)$ and consider the following problem

$$\begin{cases} -\Delta v = \lambda f(v+1), \ \ v > 0, & \text{in } \Omega \\ \quad v = 0 & \text{on } \partial\Omega. \end{cases} \tag{1.9}$$

Since $0 < \lambda < \lambda_1(f)$ we have that the functional

$$J(v) = \frac{1}{2}\int_\Omega |\nabla v|^2 - \frac{\lambda}{2}\int_\Omega f(x)v^2 - \lambda \int_\Omega fv$$

is well defined in $W_0^{1,2}(\Omega)$ and, moreover:

(1) J is coercive, indeed,

$$J(v) \ge \left(\frac{1}{2} - \frac{\lambda}{\lambda_1(f)}(\frac{1}{2}+\varepsilon)\right)\int_\Omega |\nabla v|^2 - C(\varepsilon)\lambda \int_\Omega f$$

and if $0 < \varepsilon < \frac{1}{4}(\lambda_1(f) - \lambda)$, then $\delta = \left(\frac{1}{2} - \frac{\lambda}{\lambda_1(f)}(\frac{1}{2}+\varepsilon)\right) > 0$.

(2) It is easy to see that J is Frechet-differentiable in $W_0^{1,2}(\Omega)$ and then by the *Ekeland Variational Principle* (see [64]), we obtain a sequence $\{v_k\}_{k\in\mathbb{N}} \subset W_0^{1,2}(\Omega)$, $v_k > 0$, such that

$$(i)\ v_k \rightharpoonup v \text{ weakly in } W_0^{1,2}(\Omega);\ (ii)\quad J(v_k) \to c = \inf_{w\in W_0^{1,2}(\Omega)} J(w)$$

$$\text{and}\quad (iii)\quad J'(v_k) \to 0$$

(3) As a consequence we obtain that v is a weak solution, because for all test function ϕ,

$$0 = \lim_{k\to\infty}(-\int_\Omega v_k\Delta\phi - \int_\Omega f(v_k+1)\phi) = -\int_\Omega v\Delta\phi - \int_\Omega f(v+1)\phi$$
$$= -\int_\Omega \Delta v\phi - \int_\Omega f(v+1)\phi.$$

then (1.9) has a unique positive solution $v \in W_0^{1,2}(\Omega)$. It is no too difficult to prove that in fact the convergence of the sequence is strong. Finally by setting $u = \log(v+1)$ we obtain that $u \in W_0^{1,2}(\Omega)$, $e^u - 1 \in W_0^{1,2}(\Omega)$ and

$$-\Delta u = |\nabla u|^2 + \lambda f \text{ in } \mathcal{D}'(\Omega).$$

□

Compare this kind of solvability result with the general results obtained in [85] by potential theory methods.

Remark 1.8 Notice that the following examples verifies the assertion in Theorem 1.7:

(a) If $f \in L^p(\Omega)$ with $p \geq \frac{N}{2}$, $\lambda_1(f)$ is attained by some eigenfunction $\phi_1 \in W_0^{1,2}(\Omega)$. Moreover, if $p > \frac{N}{2}$ the eigenfunction ϕ_1 is Hölder continuous and then even for $\lambda = \lambda_1(f)$ problem (1.6) has no solution. Indeed, by contradiction, if u is a solution, taking ϕ_1^2 as a test function in (1.6) we obtain that

$$2\int_\Omega \phi_1\nabla u\nabla\phi_1 = \int_\Omega |\nabla u|^2\phi_1^2 dx + \lambda_1(f)\int_\Omega f\phi_1^2 dx$$
$$= \int_\Omega |\nabla u|^2\phi_1^2 dx + \int_\Omega |\nabla\phi_1|^2 dx.$$

Therefore we obtain that

$$\int_\Omega |\nabla\phi_1 - \phi_1\nabla u|^2 dx = 0.$$

Hence $\nabla u = \dfrac{\nabla\phi_1}{\phi_1} = \nabla(\log(\phi_1))$, a contradiction with the fact that $u \in W_0^{1,2}(\Omega)$.

(b) If $f(x) = \dfrac{1}{|x|^2}$ we have the Hardy inequality. If we assume that $0 \in \Omega$ then it is well known that $\lambda_1(f) = \dfrac{(N-2)^2}{4}$ and is not attained. Then for $\lambda > \lambda_1(f)$ there is no solution and for $\lambda < \lambda_1(f)$ there exists solution.

In the case $\lambda = \lambda_1(f)$, using the *improved Hardy inequalities* (see [146] and [3]), it is possible to prove that problem (1.9) has a solution v in the space H obtained as the completion of $C_0^\infty(\Omega)$ with respect to the norm

$$||v||^2 = \int_\Omega |\nabla v|^2 dx - \lambda_1(f) \int_\Omega \frac{v^2}{|x|^2} dx.$$

As a consequence $u = \log(1 + v) \in W_0^{1,2}(\Omega)$ and $e^u - 1 \in H$. Obviously the case where $0 \notin \overline{\Omega}$ is included in the previous case a).

(c) If $f(x) = \dfrac{1}{\delta(x)}$, where $\delta(x)$ is the distance to the boundary, is not in L^1 but we have a Hardy inequality and that $f \in W^{-1,2}(\Omega)$ (see [46]). Then a slight modification of the argument in Theorem 1.7 allow us to conclude the same result.

Remark 1.9 The above nonexistence result can be easily extended to a large class of elliptic problems like

$$-\operatorname{div}(a(x, u, \nabla u)) = b(x, u, \nabla u) + \lambda f, \quad u \in W_0^{1,p}(\Omega) \tag{1.10}$$

where f and b are positive functions and

1. $|a(x, u, \xi)| \le c_1 |\xi|^{p-1}$.
2. $\mu_1 |\xi|^p \le \langle a(x, u, \xi), \xi \rangle \le \mu_2 |\xi|^p$ for all $\xi \in \mathbb{R}^N$.
3. $b(x, u, \xi) \ge c_2 |\xi|^p$.

Assume that $f \in L^1(\Omega)$ is a non-negative function, and consider

$$\Lambda(f) = \inf_{\phi \in W_0^{1,p}(\Omega)} \frac{\displaystyle\int_\Omega |\nabla \phi|^p dx}{\displaystyle\int_\Omega f |\phi|^p dx} \ge 0.$$

1. If $\Lambda(f) = 0$ then (1.10) has not solution.
2. If $\Lambda(f) > 0$ then there exists $\Lambda^* > 0$ such that problem (1.10) has no solution if $\lambda > \Lambda^*$.

Indeed, if $\phi \in C_0^\infty(\Omega)$, $\phi \ge 0$ we consider ϕ^p as test function in (1.10) and by the structural hypotheses of the equation, (1), (2) and (3), we find

$$\lambda \int_\Omega f \phi^p + c_2 \int_\Omega |\nabla u|^p \phi^p \le p c_1 \int_\Omega |\nabla u|^{p-1} \phi^{p-1} |\nabla \phi|,$$

then

$$\lambda \int_\Omega f\phi^p + c_2 \int_\Omega |\nabla u|^p \phi^p \le \varepsilon \int_\Omega |\nabla u|^p \phi^p + C(\varepsilon, p) \int_\Omega |\nabla \phi|^p$$

for ε small enough we obtain

$$\lambda \int_\Omega f\phi^p \le C(\varepsilon, p) \int_\Omega |\nabla \phi|^p,$$

and then for λ large we have a contradiction with the definition of $\Lambda(f)$.

1.2.2 Regularity

We have found the solution such that $(e^{|u|} - 1) \in W_0^{1,2}(\Omega)$. Following the examples in [52] and [69] we will discuss in the next section the existence of weaker solutions which still belong to $W_0^{1,2}(\Omega)$. With this goal in mind we will show in this section that *every* solution $u \in W_0^{1,2}(\Omega)$ of problem (1.6), and not just the *regular* one given by Theorem 1.5, enjoy *some* exponential regularity. Precisely we have the following Theorem.

Theorem 1.10 *Assume that $u \in W_0^{1,2}(\Omega)$ is a solution of problem (1.6), where $f(x) \in L^1(\Omega)$ satisfies $f(x) \ge 0$ a.e. in Ω. Then*

$$e^{\delta|u|} - 1 \in W_0^{1,2}(\Omega), \qquad \textit{for every } \delta < \frac{1}{2}. \tag{1.11}$$

Proof Assume u is a weak solution to problem (2.15) and consider as test function

$$v_\varepsilon(x) = e^{\frac{2\delta u}{1+\varepsilon u}} - 1 \in W_0^{1,2}(\Omega) \cap L^\infty(\Omega).$$

Then

$$\int\limits_\Omega |\nabla u|^2 e^{\frac{2\delta u}{1+\varepsilon u}} \frac{2\delta}{(1+\varepsilon u)^2} dx = \int\limits_\Omega |\nabla u|^2 \big(e^{\frac{2\delta u}{1+\varepsilon u}} - 1\big) + \int\limits_\Omega f\big(e^{\frac{2\delta u}{1+\varepsilon u}} - 1\big)$$

$$\ge \int\limits_\Omega |\nabla u|^2 \big(e^{\frac{2\delta u}{1+\varepsilon u}} - 1\big).$$

Therefore

$$\int\limits_\Omega |\nabla u|^2 \ge \int\limits_\Omega e^{\frac{2\delta u}{1+\varepsilon u}} \big(1 - \frac{2\delta}{(1+\varepsilon u)^2}\big) |\nabla u|^2 dx$$

If $\delta < \frac{1}{2}$ then

$$1 - \frac{2\delta}{(1+\varepsilon u)^2} > 0$$

and by Fatou's Lemma we reach

$$\int_\Omega |\nabla u|^2 \geq \frac{(1-2\delta)}{\delta^2} \int_\Omega |\nabla (e^{\delta u} - 1)|^2 dx$$

□

Remark 1.11 Notice that the regularity given by the previous theorem is optimal. Indeed if we consider $f = 0$ and $\Omega = B_1(0)$, the unit ball, then the equation admits the following family of solutions (see [69])

$$u_m(x) = \log\left(\frac{|x|^{2-N} - m}{1-m}\right), \quad 0 \leq m < 1,$$

which satisfies (2.16), but $e^{\frac{u}{2}} - 1 \notin W_0^{1,2}(\Omega)$.

The information obtained for the summability of every solution will be the key to find weaker solutions.

If f changes sign, we must require that its negative part is in $L^{\frac{N}{2}}(\Omega)$, and the regularity of u will depend on the norm of f_- in this space. More precisely we have the following result.

Theorem 1.12 *Assume that $u \in W_0^{1,2}(\Omega)$ is a solution of problem (1.6), where $f_+(x) \in L^1(\Omega)$ and $f_-(x) \in L^{\frac{N}{2}}(\Omega)$. Then*

$$e^{\delta|u|} - 1 \in W_0^{1,2}(\Omega)\,, \qquad \text{for every } \delta \text{ such that } 0 < \delta < \delta_0 = \frac{1}{1 + \sqrt{1 + S\|f_-\|_{\frac{N}{2}}}}, \tag{1.12}$$

where $S = S(N)$ is the best constant in the Sobolev inequality.

See for the proof [4].

Remark 1.13 Note that δ_0 goes to $\frac{1}{2}$ when $||f_-||_{\frac{N}{2}} \to 0$ and $\delta_0 \to 0$ when $||f_-||_{\frac{N}{2}}$ increases.

1.3 *Existence of Weaker Solutions: Connection with Elliptic Problems with Measure Data*

In this section we will show a tight relation between problems with first order quadratic terms and linear equations with measure data. This relation will imply the very strong form of non-uniqueness for distributional solutions of problem (1.6) announced above.

We recall that, given a Radon measure μ on Ω and a Borel set $E \subset \Omega$, then μ is said to be concentrated on E if $\mu(B) = \mu(B \cap E)$ for every Borel set B.

Let us also recall the *newtonian capacity* meaning. The significance of the concept of capacity for potential theory is contained in the fact that in many circumstances sets of zero capacity and only these can appear as sets of singularities for certain classes of functions or as null-sets which can be ignored.

Definition 1.14

1. If $K \subset \Omega$ is a compact set, we define $\text{cap}(K) = \text{cap}_{1,2}(K)$, the capacity of K, by

$$\text{cap}_{1,2}(K) = \inf\{ \int_\Omega |\nabla u|^2 dx \ | \ u \in \mathcal{C}_0^\infty(\Omega) \ u \geq 1 \text{ in } K \}.$$

2. If U is an open subset of Ω, we define

$$\text{cap}_{1,2}(U) = \sup\{\text{cap}_{1,2}(K) \mid K \subset U \text{ compact set}\}.$$

3. Finally if $E \subset \Omega$,

$$\text{cap}_{1,2}(E) = \inf\{\text{cap}_{1,2}(U) \mid E \subset U, \ \text{ open set}\}.$$

Notice that $\text{cap}_{1,2}$ is associated to the Newtonian potential $|x|^{2-N}$ if $N \geq 3$ and it is an outer measure on the Borel algebra. See [138] where is showed the correctness of the functional setting, and [68] and [107] for details about properties.

We define

$$T_k(s) = \max\{\min\{k, s\}, -k\}, \qquad \text{for } k > 0 \tag{1.13}$$

We call $\mathcal{M}_c(\Omega)$ the class of measures absolutely continuous with respect to the capacity, that is, if $\mu \in \mathcal{M}_c(\Omega)$ and $\text{cap}_{1,2}(E) = 0$ then $\mu(E) = 0$.

The following representation result can be found in [41] (see too [75]).

Proposition 1.15 *Assume μ a bounded radon measure in Ω. Then the following sentences are equivalents*

1. $\mu \in \mathcal{M}_c(\Omega)$.
2. $\mu \in L^1(\Omega) \cap W^{-1,2}(\Omega)$.

We will use explicitly Proposition 1.15 in the proof of the next result.

Theorem 1.16 *Let $u \in W_0^{1,2}(\Omega)$ be a solution to problem (1.6), where $f \in L^1(\Omega)$ is a positive function. Consider $v = e^u - 1$, then there exists a measure μ_s, which is concentrated on a set of zero capacity, such that*

$$\begin{cases} -\Delta v = \lambda f(x)(v+1) + \mu_s \text{ in } \mathcal{D}'(\Omega) \\ \quad v \in W_0^{1,q}(\Omega) \text{ for all } q < \frac{N}{N-1} \\ T_k(v) \in W_0^{1,2}(\Omega), \qquad \log(1+v) \in W_0^{1,2}(\Omega). \end{cases} \tag{1.14}$$

Moreover μ_s can be characterized as a weak limit in the space of bounded Radon measures, as follows:

$$\mu_s = \lim_{\epsilon \to 0} |\nabla u|^2 e^{\frac{u}{1+\varepsilon u}} \left(1 - \frac{1}{(1+\varepsilon u)^2}\right). \tag{1.15}$$

Proof Since λ does not play any role, we will take $\lambda = 1$. We set $v = e^u - 1$, then by the regularity result of Theorem 1.10 and Hölder's inequality we obtain that $v \in W_0^{1,q}(\Omega)$ for all $q < \frac{N}{N-1}$. For $\varepsilon > 0$, take $e^{\frac{u}{1+\varepsilon u}} - 1 \in L^\infty(\Omega) \cap W_0^{1,2}(\Omega)$ as test function in (1.6) Then integrating by parts,

$$\int_\Omega |\nabla u|^2 dx = \int_\Omega |\nabla u|^2 e^{\frac{u}{1+\varepsilon u}} \left(1 - \frac{1}{(1+\varepsilon u)^2}\right) dx + \int_\Omega f\,(e^{\frac{u}{1+\varepsilon u}} - 1)\, dx$$

Hence

$$\int_\Omega f\,(e^{\frac{u}{1+\varepsilon u}} - 1)\, dx \le \int_\Omega |\nabla u|^2 dx$$

and then by monotone convergence we conclude that

$$\int_\Omega f\,(e^{\frac{u}{1+\varepsilon u}} - 1) \to \int_\Omega f v dx \le \int_\Omega |\nabla u|^2 dx < +\infty. \tag{1.16}$$

On the other hand again by the same argument

$$\int_\Omega |\nabla u|^2 e^{\frac{u}{1+\varepsilon u}} \left(1 - \frac{1}{(1+\varepsilon u)^2}\right) dx \le \int_\Omega |\nabla u|^2 dx \tag{1.17}$$

then, up to a subsequence,

$$|\nabla u|^2 e^{\frac{u}{1+\varepsilon u}} \left(1 - \frac{1}{(1+\varepsilon u)^2}\right) \rightharpoonup \mu_s \tag{1.18}$$

a positive Radon measure. For simplicity of notation we set

$$w_\varepsilon(x) = |\nabla u|^2 e^{\frac{u}{1+\varepsilon u}} \left(1 - \frac{1}{(1+\varepsilon u)^2}\right),$$

and we observe that

$$\int_{u\le k} w_\varepsilon\, dx \to 0 \quad \text{for every } k \ge 0. \tag{1.19}$$

Roughly speaking we could say that μ_s is concentrated on the set $A \equiv \{x \in \Omega : u(x) = +\infty\}$. Let show that μ_s is a singular measure with respect to the classical capacity. By using the Proposition 1.15 we decompose μ_s as follows,

$$\mu_s = \mu_{s1} + \mu_{s2} \text{ where } \mu_{s1} \in L^1(\Omega) + W^{-1,2}(\Omega)$$
$$\text{and } \mu_{s2} \text{ is singular with respect to the capacity.}$$

To prove that μ_s is singular with respect to the classical $\mathrm{cap}_{1,2}$ capacity we have to show that $\mu_{s1} \equiv 0$. From Definition 2.21 in [61], for all $h \in W^{1,\infty}(\mathbb{R})$ having compact support in $\mathbb{R}$, and for all $\phi \in W_0^{1,2}(\Omega) \cap L^\infty(\Omega)$, we have in particular

$$\begin{aligned}\int_\Omega \nabla v \cdot \nabla(h(v)\phi)\, dx &= \int_\Omega |\nabla v|^2 h'(v)\, \phi\, dx + \int_\Omega h(v)\nabla v \cdot \nabla\phi\, dx\\ &= \lambda \int_\Omega f\,(v+1)\, h(v)\phi\, dx + \int_\Omega h(v)\, \phi\, d\mu_{1s}.\end{aligned} \tag{1.20}$$

On the other hand, recalling that h has compact support, one has

$$\begin{aligned}\int_\Omega \nabla v \cdot \nabla(h(v)\phi)\, dx &= \lim_{\varepsilon\to 0} \int_\Omega \nabla v_\varepsilon \cdot \nabla(h(v)\phi)\, dx\\ &= \lim_{\varepsilon\to 0} \left(\lambda \int_\Omega f\, e^{\frac{u}{1+\varepsilon u}} h(v)\, \phi\, dx + \int_\Omega w_\varepsilon\, h(v)\, \phi\, dx\right)\\ &= \lambda \int_\Omega f\,(v+1)\, h(v)\, \phi\, dx + \lim_{\varepsilon\to 0} \int_{v\le M} w_\varepsilon\, h(v)\, \phi\, dx.\end{aligned} \tag{1.21}$$

Since $v \le M$, then $u \le \log(1+M) = M_1$, therefore recalling (1.19) we conclude that

$$\left|\int_{v\le M} w_\varepsilon\, h(v)\, \phi\, dx\right| = \left|\int_{u\le M_1} w_\varepsilon\, h(v)\, \phi\, dx\right| \le c \int_{u\le M_1} w_\varepsilon\, dx \to 0 \quad \text{as } \varepsilon \to 0.$$

Hence (1.21) implies

$$\int_\Omega \nabla v \cdot \nabla(h(v)\phi)\, dx = \lambda \int_\Omega f\,(v+1)\, h(v)\, \phi\, dx,$$

which, compared with (1.20), gives

$$\int_\Omega h(v)\,\phi\,d\mu_{1s} = 0.$$

Since h and ϕ are arbitrary, we conclude that $\mu_{1s} = 0$ and then μ_s is singular. □

Remark 1.17 Notice that in the case where $e^{|u|/2} - 1 \in W_0^{1,2}(\Omega)$, that is, the regular solution, the limit in (1.15) is zero, by Lebesgue's convergence theorem.

Remark 1.18 We emphasize the fact that, given the special elliptic operator under consideration (the Laplace operator), then for measure data the notions of solution in the sense of distributions, in the sense of duality (see [138]) and of renormalized solutions (see [114] and [61]) all coincide (see also [131]).

The reader could see [33] to study elliptic problem with measures data.

Theorem 1.19 *Let $f(x)$ be a positive function in $L^r(\Omega)$, with $r > N/2$, and set*

$$\lambda_1(f) = \inf_{\phi\in W_0^{1,2}(\Omega)\setminus\{0\}} \frac{\displaystyle\int_\Omega |\nabla\phi|^2 dx}{\displaystyle\int_\Omega f\phi^2 dx}. \tag{1.22}$$

Let μ be a positive Radon measure with bounded total variation. Then, for all $\lambda < \lambda_1(f)$, problem

$$\begin{cases} -\Delta v = \lambda f(x)(v+1) + \mu \ \mathit{in}\ \mathcal{D}'(\Omega) \\ v \in W_0^{1,q}(\Omega)\ \mathit{for\ all}\ q < \frac{N}{N-1} \\ T_k(v) \in W_0^{1,2}(\Omega), \qquad \log(1+v) \in W_0^{1,2}(\Omega). \end{cases} \tag{1.23}$$

has a unique positive solution v.

Proof By the regularity of f the infimum in (1.22) is attained. We follow an approximation argument, as in [120]. Let $\{g_n\}$ a sequence of a positive bounded functions such that $g_n \to \mu$ in $\mathcal{M}_0(\Omega)$ and consider the problem

$$\begin{cases} -\Delta v_n = \lambda f(x)(v_n + 1) + g_n & \text{in } \Omega \\ v_n = 0 & \text{on } \partial\Omega. \end{cases} \tag{1.24}$$

Since $\lambda < \lambda_1(f)$, then problem (1.24) has a unique positive solution $v_n \in W_0^{1,2}(\Omega)$.

We claim that v_n is bounded in $L^{r'}(\Omega)$, where $r' = \dfrac{r}{r-1}$. If not, we can extract a subsequence (still denoted by $\{v_n\}$) such that $\|v_n\|_{r'} \to +\infty$. Then we set

$$w_n = \frac{v_n}{\|v_n\|_{r'}}.$$

Then w_n solves the equation

$$- \Delta w_n = \lambda f\, w_n + \frac{\lambda f + g_n}{\|v_n\|_{r'}}\,. \tag{1.25}$$

Since the right-hand side of (1.25) is bounded in $L^1(\Omega)$, it follows (see [138]) that w_n is bounded in $W^{1,q}(\Omega)$ for every $q < \frac{N}{N-1}$, and in $L^s(\Omega)$ for every $s < \frac{N}{N-2}$. Then one can extract a subsequence which converges weakly in the same spaces to w. Passing to the limit in (1.25), one sees that w solves

$$- \Delta w = \lambda f\, w\,. \tag{1.26}$$

Moreover, by Rellich's compactness theorem, $w_n \to w$ strongly in $L^{r'}(\Omega)$, therefore $w \neq 0$. Moreover, by a bootstrap argument applied to problem (1.26), one can check that $w \in W_0^{1,2}(\Omega)$. Therefore λ must be an eigenvalue of problem (1.26), which contradicts the assumption on λ. This proves that v_n is bounded in $L^{r'}(\Omega)$.

Therefore, by applying the same arguments with the sequence $\{w_n\}$ replaced by $\{v_n\}$, one can extract a subsequence which converges weakly to a solution v of (1.23). Since $\lambda < \lambda_1(f)$, it is easy to prove that $v(x) > 0$ in Ω. Notice that $v \in W_0^{1,q}(\Omega)$ for all $q < \frac{N}{N-1}$ and (see again [138]) $T_k(v) \in W_0^{1,2}(\Omega)$ for all $k > 0$. We prove now that $\log(v+1) \in W_0^{1,2}(\Omega)$. By using $z_n = \dfrac{v_n}{v_n + 1}$ as a test function in (1.24) we obtain that

$$\int_\Omega \frac{|\nabla v_n|^2}{(v_n+1)^2} dx = \int_\Omega f\, v_n\, dx + \int_\Omega g_n \frac{v_n}{v_n+1} dx.$$

Hence we conclude that

$$\int_\Omega \frac{|\nabla v_n|^2}{(v_n+1)^2} dx \le C.$$

Therefore by Fatou lemma we conclude that

$$\int_\Omega |\nabla \log(v+1)|^2 dx = \int_\Omega \frac{|\nabla v|^2}{(v+1)^2} dx \le C.$$

The uniqueness follows by a standard bootstrap argument. □

Remark 1.20 Note that since the principal part is the Laplacian, the well-known counterexamples to uniqueness in the space $W_0^{1,q}(\Omega)$ (see [136]) do not appear.

As a consequence we obtain the next result.

Theorem 1.21 *Let μ_s be a bounded positive measure which is concentrated on a set of zero capacity and f is in the hypothesis of Theorem 1.19. For $\lambda < \lambda_1(f)$, let v*

be the solution to problem

$$\begin{cases} -\Delta v = \lambda f(x)(v+1) + \mu_s \text{ in } \mathcal{D}'(\Omega) \\ \quad v \in W_0^{1,q}(\Omega) \text{ for all } q < \frac{N}{N-1} \\ T_k(v) \in W_0^{1,2}(\Omega), \qquad \log(1+v) \in W_0^{1,2}(\Omega). \end{cases} \tag{1.27}$$

We set $u = \log(v+1)$, *then* u *verifies*

$$\begin{cases} -\Delta u = |\nabla u|^2 + \lambda f(x) \text{ in } \mathcal{D}'(\Omega) \\ \quad u \in W_0^{1,2}(\Omega). \end{cases} \tag{1.28}$$

Proof The existence of v is obtained in Theorem 1.19, where it is also proved that $u = \log(v+1) \in W_0^{1,2}(\Omega)$. Let $\{g_n\}$ be a sequence of a bounded positive function such that $\|g_n\|_1 \le c$ and $g_n \to \mu_s$ in $\mathcal{M}_0(\Omega)$. Let v_n be the unique solution to problem

$$\begin{cases} -\Delta v_n = \lambda T_n(f(v+1)) + g_n(x) \text{ in } \Omega \\ \quad v_n \in W_0^{1,2}(\Omega). \end{cases} \tag{1.29}$$

Notice that $v_n \to v$ in $W_0^{1,q}(\Omega)$ for all $q < \frac{N}{N-1}$. We set $u_n = \log(1+v_n)$, then by a direct computation one can obtain that

$$-\Delta u_n = |\nabla u_n|^2 + \lambda \frac{T_n(f(v+1))}{v_n+1} + \frac{g_n}{v_n+1} \text{ in } \mathcal{D}'(\Omega). \tag{1.30}$$

We will show that the right-hand side of (1.30) converges to $|\nabla u|^2 + \lambda f$ in $\mathcal{D}'(\Omega)$. This will suffice to prove that u solves (2.37). It is easy to check that $\dfrac{T_n(f(v+1))}{v_n+1} \to f(x)$ in $L^1(\Omega)$.

We now claim that $\dfrac{g_n}{v_n+1} \to 0$ in $\mathcal{D}'(\Omega)$.

To prove the claim, let $A \subset \Omega$ be such that $\text{cap}(A) = 0$ and μ is concentrated on A, then for all $\varepsilon > 0$ there exists an open set U_ε such that $A \subset U_\varepsilon$ and $\text{cap}(U_\varepsilon) \le \varepsilon$. Namely for all $\varepsilon > 0$ there exists $\phi \in C_0^\infty(\Omega)$ such that $\phi \ge 0$, $\phi \equiv 1$ in U_ε and $\int_\Omega |\nabla \phi|^2 dx \le 2\varepsilon$. By using Picone inequality in Theorem (1.1), see [1], we have

$$\int_\Omega |\nabla \phi|^2 dx \ge \int_\Omega \frac{-\Delta(v_n+1)}{v_n+1} \phi^2 dx \ge \int_{U_\varepsilon} \frac{g_n}{v_n+1} dx.$$

Hence we conclude that

$$\int_{U_\varepsilon} \frac{g_n}{v_n+1} dx \le 2\varepsilon$$

for every n. Let $\phi \in \mathcal{C}_0^\infty(\Omega)$; we wish to show that

$$\lim_{n\to\infty} \int_\Omega \phi \frac{g_n}{v_n+1}\, dx = 0.$$

we can write

$$\int_\Omega \phi \frac{g_n}{v_n+1}\, dx = \int_{U_\varepsilon} \phi \frac{g_n}{v_n+1}\, dx + \int_{\Omega\setminus U_\varepsilon} \phi \frac{g_n}{v_n+1}\, dx.$$

Hence

$$|\int_\Omega \phi \frac{g_n}{v_n+1}\, dx| \le ||\phi||_\infty \int_{U_\varepsilon} \frac{g_n}{v_n+1}\, dx + \int_{\Omega\setminus U_\varepsilon} |\phi|\, g_n\, dx \le 2\varepsilon ||\phi||_\infty + \int_{\Omega\setminus U_\varepsilon} |\phi|\, g_n\, dx.$$

Now since $g_n \to \mu_s$ in $\mathcal{M}_0(\Omega)$ and μ is concentrated on $A \subset U_\varepsilon$, we conclude that

$$\int_{\Omega\setminus U_\varepsilon} |\phi|\, g_n\, dx \to 0 \text{ as } n \to \infty,$$

hence the claim follows.

To conclude the proof, let us show that

$$|\nabla u_n|^2 \to |\nabla u|^2 \quad \text{strongly in } L^1(\Omega),$$

that is,

$$\frac{|\nabla v_n|^2}{(1+v_n)^2} \to \frac{|\nabla v|^2}{(1+v)^2} \quad \text{strongly in } L^1(\Omega).$$

We recall that the well known results about the Dirichlet problem with a measure as source term we know that

$$|\nabla T_k(v_n)|^2 \to |\nabla T_k(v)|^2 \text{ as } n \to \infty, \text{ in } L^1(\Omega), \text{ for all } k > 0.$$

See [61], the appendix in [27] or the nice lectures notes by Orsina [121].

Since the sequence converges a.e. in Ω, by Vitali's theorem we only have to show that it is equi-integrable. Let $E \subset \Omega$ be a measurable set.

Then, for every $\delta \in (0, 1)$ and $k > 0$,

$$\int_E \frac{|\nabla v_n|^2}{(1+v_n)^2}\,dx = \int_{E\cap\{v_n\le k\}} \frac{|\nabla v_n|^2}{(1+v_n)^2}\,dx + \int_{E\cap\{v_n>k\}} \frac{|\nabla v_n|^2}{(1+v_n)^2}\,dx$$

$$\le \int_E |\nabla T_k(v_n)|^2\,dx + \frac{1}{(1+k)^{1-\delta}} \int_\Omega \frac{|\nabla v_n|^2}{(1+v_n)^{1+\delta}}\,dx\,.$$

The last integral is uniformly bounded with respect to n (see, for instance [34]), therefore the corresponding term can be made small by choosing k large enough. Moreover, for every $k > 0$, one has that $T_k(v_n) \to T_k(v)$ strongly on $W_0^{1,2}(\Omega)$, therefore the integral $\int_E |\nabla T_k(v_n)|^2\,dx$ is uniformly small if meas$\,(E)$ is small enough. The equi-integrability of $|\nabla u_n|^2$ follows immediately, and the proof is completed. □

Remark 1.22 If one takes the solution v to problem (2.36), and makes the change of variable $u = \log(1+v)$, then it is easy to check that u formally satisfies the equation

$$-\Delta u = |\nabla u|^2 + f + \frac{\mu_s}{1+v}\,.$$

The proof of Theorem 1.21 shows that the fraction $\frac{\mu_s}{1+v}$ is zero, which corresponds to saying that $v(x) = +\infty$ *on the set on which the singular measure* μ_s *is concentrated,* a result which is obvious in the case where μ_s is a Dirac delta concentrated on some point of Ω. For results on the behavior of solutions of elliptic equations with measure data, one should check the papers [61] and [115].

1.4 Further Results

Consider problem

$$\begin{cases} -\Delta u = \beta(u)|\nabla u|^2 + \lambda f(x) & \text{in } \Omega \\ u = 0 & \text{on } \partial\Omega, \end{cases} \tag{1.31}$$

where $\Omega \subset \mathbb{R}^N$ is a bounded domain, $f \in L^r(\Omega)$, with $r > \frac{N}{2}$, and

$$\beta : [0, +\infty) \longrightarrow [0, +\infty)$$

is a continuous nondecreasing function such that

$$\lim_{t\to+\infty} \beta(t) = +\infty. \tag{1.32}$$

We will perform a change in the dependent variable in such a way that the problem becomes semi-linear. We set

$$\gamma(t) = \int_0^t \beta(s)ds, \qquad \Psi(t) = \int_0^t e^{\gamma(s)}ds, \tag{1.33}$$

then we define

$$v(x) = \Psi(u(x)).$$

Then problem (1.31) becomes

$$\begin{cases} -\Delta v = \lambda f(x)(1 + g(v)) & \text{in } \Omega \\ v = 0 & \text{on } \partial\Omega, \end{cases} \tag{1.34}$$

where

$$g(t) = e^{\gamma(\Psi^{-1}(t))} - 1 = \int_0^t \beta(\Psi^{-1}(s))ds\,. \tag{1.35}$$

The main properties of the differentiable function $g : [0, +\infty) \longrightarrow [0, +\infty)$ are:

1. $g(0) = 0$, and g is increasing and convex
2. $\lim\limits_{s\to 0} \dfrac{g(s)}{s} = g'(0) = \beta(0)$
3. $\lim\limits_{s\to +\infty} \dfrac{g(s)}{s} = +\infty$, that is, g is superlinear at infinity.
4. But nevertheless

$$\int_0^{+\infty} \frac{ds}{1 + g(s)} = +\infty. \tag{1.36}$$

Indeed

$$\int_0^{+\infty} \frac{ds}{1+g(s)} = \int_0^{+\infty} \frac{ds}{e^{\gamma(\psi^{-1}(s))}} = \int_0^{+\infty} \frac{e^{\gamma(t)}}{e^{\gamma(t)}}dt = +\infty.$$

Notices that this last property means that g is *slightly superlinear* at infinity in the sense of the divergence of the integral in (1.36).

Now the threshold of regularity for the solution of problem (1.31) is given by $\Psi(u) \in W_0^{1,2}(\Omega)$. Without any more condition on g we can prove the following result.

Proposition 1.23 *Assume that g verifies the assumptions above. There exists λ_0 such that for $\lambda \le \lambda_0$, problem* (1.34) *has at least a positive solution $v \in W_0^{1,2}(\Omega) \cap L^\infty(\Omega)$, and then $u = \Psi^{-1}(v) \in W_0^{1,2}(\Omega)$ is a positive solution of* (1.31).

Proof We look for a super-solution in the form $\overline{v} = tw$, where w is the solution to problem

$$\begin{cases} -\Delta w = f & \text{in } \Omega \\ w \in W_0^{1,2}(\Omega) . \end{cases}$$

Notice that the function

$$h(t) = \frac{t}{1 + g(t||w||_\infty)}$$

admits a positive maximum in $\mathbb{R}^+$. If $0 < \lambda \leq \lambda_0 = \max_{\mathbb{R}^+} h(t)$, fixed t such that $t \geq \lambda(1 + g(t||w||_\infty))$, then since g is increasing

$$-\Delta\overline{v} = tf \geq \lambda f\,(1 + g(t||w||_\infty) \geq \lambda f\,(1 + g(\overline{v})).$$

To have a sub-solution we consider $\underline{v} = t_1\phi_1$ where ϕ_1 is the normalized positive eigenfunctions corresponding to the first eigenvalue to problem

$$\lambda_1(f) = \inf_{\phi \in W_0^{1,2}(\Omega)\setminus\{0\}} \frac{\int_\Omega |\nabla\phi|^2\,dx}{\int_\Omega f\,\phi^2\,dx} \tag{1.37}$$

Since $g > 0$, it suffices to have $\lambda_1(f)t_1||\phi_1||_\infty \leq \lambda$ then

$$-\Delta\underline{v} = t_1\lambda_1(f)f\,\phi_1 \leq \lambda f\,(1 + g(\underline{v}))$$

Moreover $\underline{v} \leq \overline{v}$ for t_1 small enough, by Hopf's Lemma. The result is a consequence of the usual iteration argument. □

Theorem 1.24 *There exists* $\Lambda > 0$ *such that, if* $\lambda > \Lambda$, *then problem* (1.34) *has no positive solution* $v \in W_0^{1,2}(\Omega)$.

Proof Using the properties of g, there exists a positive constant $c > 0$ such that $g(s) \geq cs - 1$. Consider now ϕ_1 a positive eigenfunction corresponding to the eigenvalue defined in (1.37); then multiplying Eq. (1.34) by ϕ_1 and using the hypothesis on g we obtain that

$$\lambda_1(f)\int_\Omega f v\phi_1\,dx = \lambda\int_\Omega f\,(g(v) + 1)\,\phi_1\,dx \geq \lambda c\int_\Omega f v\phi_1\,dx\,.$$

Hence we conclude that

$$\lambda_1(f)\int_\Omega f v\phi_1\,dx \geq \lambda c\int_\Omega f v\phi_1\,dx\,.$$

Choosing λ such that $c\lambda > \lambda_1(f)$ we obtain that $\int_\Omega f v \phi_1\, dx = 0$; therefore the strong maximum principle implies $v \equiv 0$. Hence problem (1.34) has no positive solution for $\lambda > \Lambda = \dfrac{\lambda_1(f)}{c}$. □

Corollary 1.25 *Let Λ be as in Theorem 1.24, then for $\lambda > \Lambda$ problem* (1.31) *has no solution u such that $\Psi(u) \in W_0^{1,2}(\Omega)$.*

We will see in Sect. 1.4.1, Proposition 1.37 that the nonexistence result for λ large remain true even in the distributional framework.

Remark 1.26 If, on the contrary, β is a decreasing function, it is easy to conclude that $\dfrac{g(s)}{s}$ is also decreasing. In this case problem (1.34) has a unique solution for λ small enough. The existence can be proved as in Proposition 2.19, while for uniqueness we refer to [16]. If, moreover, $\beta(s) \downarrow 0$, then $\dfrac{g(s)}{s} \downarrow 0$ as $s \to \infty$ then there exist a unique solution for all $\lambda \in \mathbb{R}^+$. These observations motivates the hypotheses of β nondecreasing to have two solutions to problem (1.34).

Next we will prove the existence of a second positive solution $w \in W_0^{1,2}(\Omega) \cap L^\infty(\Omega)$ under the following extra hypotheses on β and f. We assume that β satisfies

(H)
$$\lim_{t\to+\infty} \frac{\beta(t)}{e^{a\int_0^t \beta(s)ds}} = 0, \ \text{for some } a < \frac{4}{N+2}$$

or its equivalent form

$$\lim_{t\to+\infty} \frac{g'(t)}{(1+g(t))^a} = 0, \ \text{for some } a < \frac{4}{N+2}$$

then, using the expression of g and De L'Hôpital's rule, it is easy to check that

$$\lim_{t\to+\infty} \frac{g(t)}{t^q} = 0, \ 1 - \frac{1}{q} = a. \tag{1.38}$$

By direct calculation we check that condition **(H)** is satisfied for the elementary functions such as $\beta(s) = (\log(1+s))^\alpha$, $\beta(s) = s^\alpha$, $\beta(s) = e^s$, $\beta(s) = e^{e^s}$, etc.

Notice that in this way $q < \frac{N+2}{N-2} = 2^* - 1$, and problem (1.34) becomes variational in nature. Moreover this variational problem has a subcritical concave-convex structure. We will look for positive solutions to problem (1.34) as critical points of the associated energy functional

$$J_\lambda(u) = \frac{1}{2}\int_\Omega |\nabla u|^2 dx - \lambda \int_\Omega f\, u_+ dx - \lambda \int_\Omega f\, G(u_+) dx, \tag{1.39}$$

where

$$G(s) = \int_0^s g(t)dt,$$

which is well defined in $W_0^{1,2}(\Omega)$.

Remark 1.27

1. Notice that since the nonlinear term $g(u)$ has slightly super-linear growth, in general, doesn't verify Ambrosetti-Rabinowitz assumption ensuring that all Palais-Smale sequences for the associated energy functional are bounded. Namely we must to prove the boundedness of a Palais-Smale sequence with different arguments.
2. The result of Theorem 1.31 is true if we assume that f satisfies hypothesis.

$$\textbf{(F)} \qquad f(x) \in L^r(\Omega), \qquad \text{for } r > \frac{2^*}{2^* - (q+1)},$$

where q is defined by (1.38).

For simplicity we will consider the case where f it is a non-negative, bounded function.

We will prove the existence of at least two positive solutions for λ small enough. To overpass the difficulty explained in the Remark 1.27 about the boundedness of a Palais-Smale sequence, we use the following general result proved in [87].

Theorem 1.28 *Let X be a Banach space endowed with the norm $||.||$ and let $I \subset \mathbb{R}^+$ be an interval. Let $\{J_\alpha\}_{\alpha \in I}$ be a family of functionals on X of the form*

$$J_\alpha(u) = A(u) - \alpha B(u)$$

where $B(u) \geq 0$ and such that $A(u)$ or $B(u) \to +\infty$ as $||u|| \to \infty$. We assume that there exist two points $v_1, v_2 \in X$ such that, setting

$$\Gamma = \{\gamma \in C([0,1]; X),\ \gamma(0) = v_1,\ \gamma(1) = v_2\},$$

there hold, for all $\alpha \in I$,

$$c(\alpha) = \inf_{\gamma \in \Gamma} \max_{t \in [0,1]} J_\alpha(\gamma(t)) > \max\{J_\alpha(v_1), J_\alpha(v_2)\}.$$

Then for almost every $\alpha \in I$, there exists a sequence $\{v_k\} \subset X$ such that : i) $\{v_k\}$ is bounded; ii) $J_\alpha(v_k) \to c(\alpha)$ and iii)$J'_\alpha(v_k) \to 0$ in X', the dual of X.

Precisely, by using the previous abstract result, we have the following result.

Theorem 1.29 *Assume that* (1.32) *and* (**H**) *hold, that* $f(x)$ *is bounded and non-negative, and that the functional* J_λ *has the geometry of the mountain pass, that is, there exist two points* $v_1, v_2 \in W_0^{1,2}(\Omega)$ *such that, setting*

$$\Gamma = \{\gamma \in C([0,1]; W_0^{1,2}),\ \gamma(0) = v_1,\ \gamma(1) = v_2\},$$

there holds

$$c(\lambda) = \inf_{\gamma\in\Gamma} \max_{t\in[0,1]} J_\lambda(\gamma(t)) > \max\{J_\lambda(v_1), J_\lambda(v_2)\}. \tag{1.40}$$

Then problem (1.34) *has a mountain-pass type positive solution u.*

Proof By a continuity argument there exists $\epsilon > 0$ such that for all $\alpha \in I = [1 - \epsilon, 1 + \epsilon]$, the family of functionals $\{J_{\lambda,\alpha}\}_{\alpha\in I}$ defined by

$$J_{\lambda,\alpha}(u) = \frac{1}{2}\int_\Omega |\nabla u|^2 dx - \lambda\alpha\Big(\int_\Omega f\,u_+ dx + \int_\Omega f\,G(u_+)dx\Big)$$

have the same geometry, in the sense that

$$c(\lambda,\alpha) = \inf_{\gamma\in\Gamma} \max_{t\in[0,1]} J_{\lambda,\alpha}(\gamma(t)) > \max\{J_{\lambda,\alpha}(v_1), J_{\lambda,\alpha}(v_2)\}.$$

That is (v_1, v_2) are independent of $\alpha \in I$.

By Theorem 1.28 we obtain that for almost every $\alpha \in I$ there exists a sequence $\{v_k^{(\alpha)}\}$ such that: *i)* $\{v_k^{(\alpha)}\}$ is bounded; *ii)* $J_{\lambda,\alpha}(v_k^{(\alpha)}) \to c(\lambda,\alpha)$ and *iii)* $J'_{\lambda,\alpha}(v_k^{(\alpha)}) \to 0$ in $W^{-1,2}(\Omega)$. Namely, we obtain a bounded Palais-Smale sequence for $J_{\lambda,\alpha}$ for almost all $\alpha \in I$. Since g verifies (**H**), then using a compactness argument we obtain that the Palais-Smale condition holds, i.e., up to a subsequence, $v_k^{(\alpha)} \to v^{(\alpha)}$ strongly in $W_0^{1,2}(\Omega)$, where $v^{(\alpha)}$ is a positive solution to problem

$$\begin{cases} -\Delta v^{(\alpha)} = \lambda\alpha f\,(1 + g(v^{(\alpha)})) & \text{in } \Omega \\ \quad v^{(\alpha)} = 0 & \text{on } \partial\Omega, \end{cases} \tag{1.41}$$

such that $J_{\lambda,\alpha}(v^{(\alpha)}) = c(\lambda,\alpha)$.

We have to prove that the conclusion in Theorem 1.28 holds for $\alpha = 1$. Let $\{\alpha_n\}$ be a decreasing sequence in I such that $\alpha_n \downarrow 1$ as $n \to \infty$ and consider $v^{(\alpha_n)}$ the corresponding solution to problem (1.41). We will prove that $\{v^{(\alpha_n)}\}$ is bounded in $W_0^{1,2}(\Omega)$.

For the simplicity of notation we set $v_n = v^{(\alpha_n)}$.

If $||v_n||_\infty \le C$ for all n, then using (1.41) and by the condition on f and g we conclude that $||v_n||_{W_0^{1,2}} \le C_1$.

Assume now that $||v_n||_\infty \to +\infty$ as $n \to \infty$. Consider ϕ_1 the positive eigenfunction associated to the first eigenvalue

$$\begin{cases} -\Delta\phi_1 = \lambda_1(f)f\phi_1 & \text{in } \Omega \\ \phi_1 = 0 & \text{on } \partial\Omega. \end{cases} \tag{1.42}$$

By taking ϕ_1 as a test function in (1.41) we obtain that

$$\lambda_1(f)\int_\Omega f\phi_1 v_n dx = \lambda\alpha_n \int_\Omega f\phi_1 + \lambda\alpha_n \int_\Omega f g(v_n)\phi_1 dx.$$

Since the hypothesis (3) on g holds, it is easy to check that there exists a constant C_1 such that

$$\int_\Omega f\phi_1 v_n dx \le C_1 \quad \text{and} \quad \int_\Omega f\phi_1 g(v_n)dx \le C_1.$$

Let now ϕ_2 be the solution to problem

$$\begin{cases} -\Delta\phi_2 = f & \text{in } \Omega \\ \phi_2 = 0 & \text{on } \partial\Omega. \end{cases} \tag{1.43}$$

Notice that, by Hopf Lemma, there exist $c_1, c_2 > 0$ such that $c_1\phi_1 \le \phi_2 \le c_2\phi_1$. Taking ϕ_2 as a test function in (1.41) we obtain that

$$\int_\Omega f v_n dx = \lambda\alpha_n \int_\Omega f\phi_2 + \lambda\alpha_n \int_\Omega f g(v_n)\phi_2 dx. \tag{1.44}$$

Since $\phi_2 \le c_2\phi_1$ we conclude that

$$\int_\Omega f v_n dx \le \lambda\alpha_n \int_\Omega f\phi_2 + c_2\lambda\alpha_n \int_\Omega f g(v_n)\phi_1 dx.$$

Hence,

$$\int_\Omega f v_n dx \le C. \tag{1.45}$$

As $J_{\lambda,\alpha_n}(v_n) = c(\lambda,\alpha_n) \le c(\lambda) + 1$, by using (1.45) we obtain that

$$\int_\Omega f\,(g(v_n)v_n - 2G(v_n))dx \le C. \tag{1.46}$$

We now prove the energy estimate. Assume by contradiction that $||v_n||_{W_0^{1,2}} \to \infty$ as $n \to \infty$.

We set $w_n = \dfrac{v_n}{||v_n||_{W_0^{1,2}}}$, then $||w_n||_{W_0^{1,2}} = 1$, hence there exists $w_0 \in W_0^{1,2}(\Omega)$ such that, up to a subsequence, $w_n \rightharpoonup w_0$ weakly in $W_0^{1,2}(\Omega)$ and $w_n \to w_0$ strongly in $L^p(\Omega)$ for all $p < \frac{2N}{N-2}$ if $N \geq 3$ (in all $p < \infty$ if $N = 1, 2$). Moreover w_n verifies

$$-\Delta w_n = \frac{\alpha_n \lambda f}{||v_n||_{W_0^{1,2}}} + \alpha_n \lambda \frac{f\, g(v_n)}{||v_n||_{W_0^{1,2}}}.$$

Since $w_n \rightharpoonup w_0$ weakly in $W_0^{1,2}$ we obtain that

$$\int_\Omega -\Delta w_0 \phi = \lim_{n\to\infty} \lambda \int_\Omega \frac{f\, g(v_n)}{||v_n||_{W_0^{1,2}}} \phi \quad \text{for all } \phi \in C_0^\infty(\Omega). \tag{1.47}$$

From (1.44) and (1.45) we obtain that $fg(v_n)$ is bounded in $L^1_{\text{loc}}(\Omega)$. Therefore (1.47) implies $w_0 = 0$. Let $z_n = t_n v_n$ where t_n is defined as

$$t_n = \inf \left\{ t \in [0, 1] \mid J_{\lambda,\alpha_n}(tv_n) = \max_{t\in[0,1]} J_{\lambda,\alpha_n}(tv_n) \right\}.$$

We prove that $t_n \in (0, 1)$ for n large enough.

That $t_n \neq 0$ is obvious because $J_{\lambda,\alpha_n}(0) = 0$ for all values of α_n. To show that $t_n \neq 1$ we claim that

$$\lim_{n\to\infty} J_{\lambda,\alpha_n}(z_n) = +\infty\,. \tag{1.48}$$

We argue by contradiction; if $\liminf_{n\to\infty} J_{\lambda,\alpha_n}(z_n) \leq M$, we set $u_n = \sqrt{4M} w_n$, then $u_n \rightharpoonup 0$ weakly in $W_0^{1,2}(\Omega)$, hence

$$\int_\Omega f\, G(u_n)dx, \int_\Omega f\, u_n dx \to 0 \text{ as } n \to \infty.$$

Therefore we obtain that

$$J_{\lambda,\alpha_n}(u_n) = 2M - \frac{\alpha_n \lambda}{2} \int_\Omega f u_n dx - \alpha_n \lambda \int_\Omega f\, G(u_n) dx \geq \frac{3}{2} M \text{ as } n \to \infty. \tag{1.49}$$

On the other hand, using the definition of z_n and since $u_n = \dfrac{\sqrt{4M}}{||v_n||_{W_0^{1,2}}} v_n$, we obtain that

$$J_{\lambda,\alpha_n}(u_n) \leq J_{\lambda,\alpha_n}(z_n) \leq M$$

a contradiction with (1.49). Hence (1.48) is proved.

Therefore, taking into account that $J_{\lambda,\alpha_n}(v_n) = c_{\lambda,\alpha_n} \le c(\lambda) + 1$ and by the claim, we conclude $t_n \ne 1$ for n large enough. As a consequence by the definition of z_n we have $\langle J'_{\lambda,\alpha_n}(z_n), z_n\rangle = 0$, hence we conclude that

$$J_{\lambda,\alpha_n}(z_n) = \frac{\alpha_n\lambda}{2}\int_\Omega f\,(g(z_n)z_n - 2G(z_n))\,dx - \frac{\alpha_n\lambda}{2}\int_\Omega f\,z_n\,dx\,.$$

Since $\displaystyle\int_\Omega f\,z_n\,dx \le \int_\Omega f\,v_n\,dx \le C$, by (1.48) we conclude that

$$\int_\Omega f\,(g(z_n)z_n - 2G(z_n))\,dx \to +\infty, \quad \text{as } n \to \infty.$$

Notice that the function $l(s) = g(s)s - 2G(s)$ is increasing then

$$g(z_n)z_n - 2G(z_n) \le g(v_n)v_n - G(v_n)$$

and therefore

$$\int_\Omega f\,(g(v_n)v_n - 2G(v_n))\,dx \to \infty \qquad \text{as } n \to \infty\,,$$

a contradiction with (1.46). As a consequence we conclude that

$$||v_n||_{W_0^{1,2}} \le C_1.$$

Therefore $v_n \to v$ weakly in $W_0^{1,2}(\Omega)$ and strongly in $L^\theta(\Omega)$ for all $\theta < \frac{2N}{N-2}$. Using again the hypotheses on g and by a simple compactness argument we obtain that v is a weak solution to problem (2.55) and then easily we obtain that $v_n \to v$ strongly in $W_0^{1,2}(\Omega)$.

Therefore we conclude that v is a non-negative solution to problem (2.55) such that

$$c(\lambda, \alpha_n) = J_{\lambda,\alpha_n}(v_n) \to J_\lambda(v) \quad \text{as } n \to \infty.$$

Hence we get the existence of a positive solution v to problem (2.55) with $J_\lambda(v) = c(\lambda)$ and the proof is complete. □

Corollary 1.30 *There exists λ_0 such that if $0 < \lambda \le \lambda_0$, then the functional J_λ has the geometry of the mountain pass and then problem* (1.34) *has at least two positive solutions.*

Moreover we also can obtain the following stronger global result in λ.

Theorem 1.31 *Under the same assumptions of Theorem 1.29, let* λ^* *be defined by*

$$\lambda^* = \sup\{\lambda \geq 0 \text{ such that problem } (1.34) \text{ has a positive solution } \}. \tag{1.50}$$

Then for all $\lambda \in (0, \lambda^*)$*, problem* (1.34) *has at least two positive solutions. If* $\lambda = \lambda^*$*, then problem* (1.34) *has at least one positive solution.*

As a direct application of Theorem 1.31 we get the following result.

Corollary 1.32 *Assume that* β *is an increasing function such that* (1.32) *and* (**H**) *hold. Let* λ^* *be defined as in* (1.50)*. Then for every* $\lambda < \lambda^*$*, problem* (1.31) *admits a least two solutions* $u_1,\ u_2$ *such that* $\Psi(u) \in W_0^{1,2}(\Omega)$ *where* Ψ *is defined as in* (1.33)*. For* $\lambda = \lambda^*$ *problem* (1.31) *admits at least one solution* u *with* $\Psi(u) \in W_0^{1,2}(\Omega)$ *and for* $\lambda > \lambda^*$ *problem* (1.31) *has no solution such that* $\Psi(u) \in W_0^{1,2}(\Omega)$*.*

We use results in [13] and in [16]. See [4] for a detailed proof of the results above.

We can also prove that no distributional solution exists for problem (1.31) for $\lambda > \lambda^*$.

1.4.1 Regularity and Existence of Weaker Solutions

We will consider the problem (1.31) where β satisfies more general hypotheses than in the previous section. Precisely we will only assume that,

(b1) β is a continuous non-negative function on $[0, +\infty)$.
(b2) $\liminf_{t\to\infty} \beta(t) \in (0, +\infty]$.

In the existence result, Theorem 1.35, we will use an extra hypothesis, that is,

$$\lim_{t\to\infty} \frac{\beta(t)}{e^{a\gamma(t)}} = 0, \qquad \begin{cases} a < \dfrac{2}{N} \text{ if } N \geq 3 \\ \\ a < 1 \text{ if } N = 1,\ 2 \end{cases} \tag{1.51}$$

or its equivalent form $\lim_{t\to\infty} \dfrac{g'(t)}{(1 + g(t))^a} = 0$. Then it is easy to check that

$$\lim_{t\to\infty} \frac{g(t)}{t^q} = 0,\ q = \frac{1}{1-a} < \frac{N}{N-2} \text{ if } N \geq 3$$

and

$$\lim_{t\to\infty} \frac{g(t)}{t^q} = 0,\ q < \infty \text{ if } N = 1,\ 2$$

(we recall that the functions γ and g have been defined at the beginning of the previous section). This condition is verified if β is any elementary function. We will also suppose that $f \in L^1(\Omega)$ is a positive function.

By a solution to problem (1.31) we mean a function $u \in W_0^{1,2}(\Omega)$ such that $\beta(u)|\nabla u|^2 \in L^1(\Omega)$ and u is a solution in distribution sense to problem

$$\begin{cases} -\Delta u = \beta(u)|\nabla u|^2 + f & \text{in } \Omega \\ u = 0 & \text{on } \partial\Omega. \end{cases} \tag{1.52}$$

Once u is fixed one can consider (1.52) as a problem with L^1 right-hand side. Notice that, in this case the renormalized solution coincide with the distributional solution (see [61] for details). As a consequence we obtain that $T_k(u) \in W_0^{1,2}(\Omega)$ for all $k > 0$. Since $\beta(t) > A > 0$ as $\tau \to \infty$ and by the fact that $\beta(u)|\nabla u|^2 \in L^1(\Omega)$ we conclude that $u \in W_0^{1,2}(\Omega)$. We start with the following regularity result.

Theorem 1.33 *Assume that $u \in W_0^{1,2}(\Omega)$ is a solution of problem* (1.31), *where $f(x) \in L^1(\Omega)$ satisfies $f(x) \geq 0$ a.e. in Ω. Then*

$$\Psi_\delta(u) \in W_0^{1,2}(\Omega) \qquad \text{for every } \delta < \frac{1}{2}, \text{ where } \Psi_\delta(s) = \int_0^s \sqrt{\beta(t)}e^{\delta\gamma(t)}dt. \tag{1.53}$$

We now consider the reverse problem, namely we have the following result.

Theorem 1.34 *Let μ_s be a bounded positive measure which is concentrated on a set with zero capacity. Let v be a solution to problem*

$$\begin{cases} -\Delta v = f(x)(1 + g(v)) + \mu_s \text{ in } \mathcal{D}'(\Omega) \\ v \in W_0^{1,q}(\Omega) \text{ for all } q < \dfrac{N}{N-1}, \\ f(x)(g(v) + 1) \in L^1(\Omega) \end{cases} \tag{1.54}$$

If we define $u = \Psi^{-1}(v)$, where Ψ is given by (1.33), *then u solves*

$$\begin{cases} -\Delta u = \beta(u)|\nabla u|^2 + f(x) \text{ in } \mathcal{D}'(\Omega) \\ u \in W_0^{1,2}(\Omega) \\ \beta(u)|\nabla u|^2 \in L^1(\Omega). \end{cases} \tag{1.55}$$

We now give a fairly general example for which problem (2.68) has a solution.

Theorem 1.35 *Assume that $f \in L^\infty(\Omega)$ and assumption* (1.51) *holds, then problem* (2.68) *has a positive solution for λ small enough depending on μ. This implies that problem* (1.31) *admits infinitely many solutions for small λ.*

For the proof we will use the following result that can be found in [19].

Theorem 1.36 *Consider the problem*

$$\begin{cases} -\Delta v = v^q + \lambda v \text{ in } \mathcal{D}'(\Omega) \\ v_{|\partial\Omega} = 0. \end{cases} \tag{1.56}$$

Assume that $q < \frac{N}{N-2}$ *if* $N \geq 3$ *or* $q < \infty$ *if* $N = 1, 2$, *then there exists* λ^* *such that problem* (1.56) *has a solution if* $\lambda < \lambda^*$.

Proof (Theorem 1.35)
Consider the problem

$$\begin{cases} -\Delta v = \lambda(v^q + C + 1) + \mu_s \text{ in } \mathcal{D}'(\Omega) \\ v_{|\partial\Omega} = 0. \end{cases} \tag{1.57}$$

By (1.51), all solutions of (1.57) are supersolutions to problem (2.68), for a suitable C. Let $v_1 = \lambda^{\frac{1}{q-1}} v$, then

$$-\Delta v_1 = v_1^q + \lambda^{\frac{1}{q-1}}(\lambda c + \mu_s). \tag{1.58}$$

there exists $\lambda_0 > 0$ such that for all $\lambda < \lambda_0$ Eq. (1.58) has a solution, hence problem (1.57) has a solution which is a supersolution to problem (2.68). Using an iteration argument we get the existence result. □

The existence of infinitely many solutions for problem (1.31) (and of (1.6)) should be compared with a uniqueness result proved by Korkut-Pašić-Žubrinić in [95]. In that article, which extends to more general operators than the Laplacian, they prove that, in the case where $\beta(s) \in L^\infty(\mathbb{R}) \cap L^1(\mathbb{R})$ and $f = 0$, the only solution $u \in W_0^{1,2}(\Omega)$ of (1.31) is zero. In the light of the change of variable used here, one can explain this uniqueness result (and also give an alternative proof in our particular framework). The above result should be also compared with the existence results by Porretta-Segura [130] in the case where $\beta(s)$ is a positive function such that $\lim_{s\to+\infty} \beta(s) = 0$. In that paper it is proved that, under this assumption, a solution of (1.31) exists for all $\lambda > 0$.

In the next Sect. 3 we will see some intermediate cases about the behavior of β.

Finally, we give a non-existence result which completes the statement of Theorem 1.24 and Corollary 1.25.

Proposition 1.37 *Assume that* β *is an nondecreasing function such that* **(H)** *and* (1.32) *hold. Let* λ^* *defined as in* (1.50), *then problem* (1.31) *admits no distributional solution for* $\lambda > \lambda^*$.

Proof By contradiction. Let $\lambda > \lambda^*$ be such that problem (1.31) has a solution. We set $v = \Psi(u)$ where Ψ is defined as in (1.33), then we obtain that v satisfies to problem

$$\begin{cases} -\Delta v = \lambda f(x)(g(v)+1) + \mu_s \text{ in } \mathcal{D}'(\Omega) \\ v \in W_0^{1,q}(\Omega) \text{ for all } q < \frac{N}{N-1}. \end{cases} \tag{1.59}$$

where μ_s is a positive measure which only charges a set with singular measure. Consider now the problem

$$\begin{cases} -\Delta w = \lambda f(x)(g(w)+1) \text{ in } \Omega \\ w_{|\partial\Omega} = 0. \end{cases} \tag{1.60}$$

Since $w_0 = 0$ is a strictly supsolution and $w_1 = v$ is a supersolution, then using an iteration argument we obtain that problem (1.60) has at least a positive solution for $\lambda > \lambda^*$, a contradiction with the definition of λ^*. Hence we conclude. □

2 Some Parabolic Problems with Critical Growth in the Gradient

2.1 Introduction

In this chapter we will consider the parabolic KPZ problem described in Sect. 1. More precisely we will study the following problem

$$\begin{cases} u_t - \Delta u = |\nabla u|^2 + f(x,t) & \text{in } Q \equiv \Omega \times (0,+\infty) \\ u(x,t) = 0 & \text{on } \partial\Omega \times (0,+\infty), \\ u(x,0) = u_0(x) & \text{in } \Omega, \end{cases} \tag{2.1}$$

where Ω is a bounded regular domain and f, u_0 are positive functions satisfying some hypotheses that we will specify later.

Existence results for problem (2.1) in the whole $\mathbb{R}^N$, with a regular data u_0 and $f \equiv 0$ is well known, we refer to [80], where the Cauchy problem for the equation

$$u_t - \Delta u = |\nabla u|^q, \quad q \geq 1 \tag{2.2}$$

is studied. We refer also to the paper [22] where problem (2.2) is studied in the case $q \leq 2$ and some quantitative properties of the solutions are obtained in that case.

To obtain an existence result for problem (2.1) in the case where the data are bounded: as in the elliptic case it suffices to use a change of unknown of the form $v = e^u - 1$ also known as Cole-Hopf transformation, to transform the equation into a linear problem, which can then be solved by super/sub-solution methods. In the case where the operator is more general or in the case where the data are unbounded this change of variable cannot be done, but it can be replaced with the use of exponential-type test function, whose role is again to get rid of the quadratic term in (2.1) (see [39, 58]). The case where the Laplace operator is replaced by a nonlinear operator like the p-Laplacian has been studied in [6, 55, 56, 58, 71, 81, 122] and references therein.

We shall consider in this chapter the problem of regularity, uniqueness and non uniqueness of solutions to problem (2.1). More precisely, we will show that

$$e^{\delta u} - 1 \in L^2(0,T; W_0^{1,2}(\Omega)) \cap C^0([0,\infty); L^2(\Omega)) \qquad \text{for all } \delta < 1/2, \text{ for all } T > 0, \tag{2.3}$$

$$\int_\Omega e^{u(x,t)}\,dx < \infty \qquad \text{for all } t \geq 0. \tag{2.4}$$

The result (2.3) resembles the corresponding the one for elliptic equations with quadratic gradient term, obtained in Sect. 1 (see [4]), and has in common with it the fact that the elliptic part of the equation is never used for the regularity, more precisely that the main estimate come from the quadratic term on the right-hand side. Moreover, as in the elliptic case, no regularity on the datum f is assumed (only $f \in L^1_{\text{loc}}(\overline{Q})$ is required). However the proof of the parabolic result is more complicated, since one has to estimate the term with the time derivative of u.

Then we proceed in performing a precise analysis in what happens in the Cole-Hopf change of variable, particularly if one does not assume that the transformed function $v = \Psi(u)$ belongs to the *energy space*, that is, $L^2(0,T; W_0^{1,2}(\Omega)) \cap C^0([0,\infty); L^2(\Omega))$, for all $T > 0$. We will show a striking non-uniqueness result and a direct correspondence between solutions of problem (2.1) and solutions of semilinear problems with measure data, that is, we consider the following linear problem

$$\begin{cases} v_t - \Delta v = f(x,t)(v+1) + \mu_s & \text{in } \mathcal{D}'(Q) \\ v(x,t) = 0 & \text{on } \partial\Omega \times (0,+\infty), \\ v(x,0) = v_0(x) & \text{in } \Omega, \end{cases} \tag{2.5}$$

where μ_s is a singular positive Radon measure. Here "singular" means that it is concentrated on a set with zero capacity, where by "capacity" we mean the parabolic capacity introduced by Pierre in [128] and studied by Droniou, Porretta and Prignet in [62].

More precisely, under appropriate integrability assumptions on the data f and v_0, we show (Theorem 2.10) that problem (2.5) admits exactly one solution, and that if we apply the change of variable $u = \log(1+v)$, then u is a solution of problem (2.1), with $\beta \equiv 1$. We could summarize this non-uniqueness result as follows:

There exists a one to one correspondence between solutions to problem (2.1) *and singular measures concentrated in zero parabolic capacity sets in the cylinder* $Q = \Omega \times [0,\infty)$.

Therefore problem (2.1) admits infinitely many solutions, whose singularities can be prescribed. The idea behind the result is very simple: if one makes formally the change of variable, then $u = \log(1 + v)$ solves the equation

$$u_t - \Delta u = |\nabla u|^2 + f + \frac{\mu_s}{1+u},$$

but if μ_s is a singular measure (for instance, if $\mu_s = \mu_s(x) = \delta_{x_0}(x)$ in space dimension $N \geq 2$), then v is infinite on the set where μ_s is concentrated, therefore the last term in Eq. (2.1) vanishes. Of course this is just a formal calculation, but the result will be justified rigorously.

An inverse result can also be proved (see Theorem 2.13): every solution u of problem (2.1) corresponds, via the change of variable $v = e^u - 1$, to the solution v of an equation of the form (2.5), for a singular measure μ_s which is determined by u. Among these infinitely many functions there is only one, which we call the *regular* one, which corresponds to $\mu_s = 0$. This function is such that $v = e^u - 1 \in L^2(0, T; W_0^{1,2}(\Omega))$, and is unique in the larger class of functions such that $e^{u/2} - 1 \in L^2(0, T; W_0^{1,2}(\Omega))$. All the other solutions only satisfy $e^{\delta u} - 1 \in L^2(0, T; W_0^{1,2}(\Omega))$ for every $\delta < 1/2$.

It is interesting to point out that we also get infinitely many solutions by singular perturbation of the initial data in the transformed problem. More precisely if v is the renormalized solution to problem

$$\begin{cases} v_t - \Delta v = 0 & \text{in } \mathcal{D}'(Q) \\ v(x,t) = 0 & \text{on } \partial\Omega \times (0, +\infty), \\ v(x,0) = \nu_s & \text{in } \Omega, \end{cases}$$

where ν_s is a singular positive measure with respect to the classical Lebesgue measure, then $u = \log(v + 1)$ solves problem (2.14) with $f \equiv 0$ and $u_0(x) \equiv 0$. We refer to Sect. 2.4.3 for more details.

The content of this chapter is a part of the article [5].

2.1.1 Preliminaries

Let Ω be a bounded domain in $\mathbb{R}^N$, $N \geq 1$. We will first precise the notation and then we will give the precise meaning of solution that appear in the chapter. Finally we collect some results for the heat equation.

We will denote by Q the cylinder $\Omega \times (0, \infty)$; moreover, for $0 < t_1 < t_2$, we will denote by Q_{t_1}, Q_{t_1,t_2} the cylinders $\Omega \times (0, t_1)$, $\Omega \times (t_1, t_2)$, respectively. The data in problem (2.1), $u_0(x)$ and $f(x, t)$, are positive functions defined in Ω, Q, respectively, such that $u_0 \in L^1(\Omega)$ and $f \in L^1(Q_T)$, for every $T > 0$.

We will consider $L^r(0,T;L^q(\Omega))$ denote the usual vectorial Lebesgue spaces, and the Lebesgue space valued in a Sobolev space, $L^r(0,T;W_0^{1,q}(\Omega))$, see for instance [63]. We recall that $W^{-1,q'}(\Omega)$ means the dual space of $W_0^{1,q}(\Omega)$, $q > 1$, where $\frac{1}{q}+\frac{1}{q'}=1$. For the sake of brevity, instead of writing $u(x,t) \in L^r(0,\tau;W_0^{1,q}(\Omega))$ for every $\tau > 0$, we shall write $u(x,t) \in L^r_{\rm loc}([0,\infty);W_0^{1,q}(\Omega))$. Similarly, we shall write $u \in L^q_{\rm loc}(\overline{Q})$ instead of $u \in L^q(Q_\tau)$ for every $\tau > 0$. The usual truncation to level $k > 0$, $T_k(s)$, is defined in (1.13). Notice that,

$$T_k(s) = \begin{cases} s \text{ if } |s| \le k \\ k\frac{s}{|s|} \text{ if } |s| > k. \end{cases}$$

As usual, if $1 \le q < N$, we will denote by $q^* = Nq/(N-q)$ its Sobolev conjugate exponent.

Classes of Solutions We precise next the meaning of *solution*.

Definition 2.1 Assume $u_0 \in L^2(\Omega), f \in L^2_{loc}([0,\infty), W^{-1,2}(\Omega))$. We say that u is a weak solution to problem

$$\begin{cases} v_t - \Delta v = f & \text{in } Q \\ v(x,t) = 0 & \text{on } \partial\Omega \times (0,+\infty), \\ v(x,0) = u_0 & \text{in } \Omega, \end{cases} \tag{2.6}$$

if $u \in L^2_{loc}([0,\infty), W_0^{1,2}(\Omega))$, $u_t \in L^2_{loc}([0,\infty), W^{-1,2}(\Omega))$ and for all $T > 0$,

$$\int_0^T\int_\Omega u_t\varphi\,dx\,dt + \int_0^T\int_Q \langle\nabla u, \nabla\varphi\rangle\,dt = \int_0^T\int_\Omega f(x,t)\varphi\,dx\,dt,$$

for any $\varphi \in L^2([0,T], W_0^{1,2}(\Omega))$.

In the case of data with weaker integrability we have to use the following weaker notion of solution.

Definition 2.2 Assume $f \in L^1(\Omega \times (0,T))$, $u_0 \in L^1(\Omega)$. We say that $u \in C([0,\infty);L^1(\Omega))$ is a distributional solution of problem (2.6), if for all $\phi \in C_0^\infty(Q)$ we have that

$$\iint_Q -\phi_t\,u\,dxdt + \iint_Q u(-\Delta\phi)\,dx\,dt = \iint_Q f\phi\,dxdt + \int_\Omega u_0(x)\phi(x,0)\,dx. \tag{2.7}$$

The meaning of solution for problem (2.1) is then as follows.

Definition 2.3 We say that $u(x,t)$ is a distributional solution to problem (2.1) if $u \in C([0,\infty); L^1(\Omega)) \cap L^2_{\text{loc}}([0,\infty); W_0^{1,2}(\Omega))$, $|\nabla u|^2 \in L^1_{\text{loc}}(\overline{Q})$, and if for all $\phi(x,t) \in C_0^\infty(Q)$ one has

$$-\iint_Q u\,\phi_t\,dx\,dt + \iint_Q \nabla u\cdot\nabla\phi\,dx\,dt = \iint_Q |\nabla u|^2\,\phi\,dx\,dt + \iint_Q f\phi\,dx\,dt$$

and

$$u(\cdot,0) = u_0(\cdot) \qquad \text{in } L^1(\Omega).$$

Remark 2.4 Note that the previous definition implies that, for every bounded, Lipschitz continuous function $h(s)$ such that $h(0) = 0$, and for every $\tau > 0$, one has

$$\int_\Omega H(u(x,\tau))\,dx - \int_\Omega H(u_0(x))\,dx + \iint_{Q_\tau} |\nabla u|^2\,h'(u)\,dx\,dt$$
$$= \iint_{Q_\tau} |\nabla u|^2\,h(u)\,dx\,dt + \iint_{Q_\tau} f\,h(u)\,dx\,dt\,,$$

where $H(s) = \displaystyle\int_0^s h(\sigma)\,d\sigma$.

Similarly, if $h(s)$ is Lipschitz continuous and bounded, if $\phi(x,t) \in L^2(0,\tau; W_0^{1,2}(\Omega)) \cap L^\infty(Q_\tau)$ and $\phi_t \in L^2(0,\tau; W^{-1,2}(\Omega))$, then one has

$$\int_\Omega H(u(x,\tau))\,\phi(x,\tau)\,dx - \int_\Omega H(u_0(x))\,\phi(x,0)\,dx$$
$$-\iint_{Q_\tau} \phi_t\,H(u)\,dx\,dt + \iint_{Q_\tau} |\nabla u|^2\,h'(u)\,\phi\,dx\,dt + \iint_{Q_\tau} \nabla u\cdot\nabla\phi\,h(u)\,dx\,dt$$
$$= \iint_{Q_\tau} |\nabla u|^2\,h(u)\,\phi\,dx\,dt + \iint_{Q_\tau} f\,h(u)\,\phi\,dx\,dt\,.$$

Classical Results for the Heat Equation The regularity and convergence results about parabolic equations with L^1 or measure data can be seen for instance in [42] and [34]. We summarize the results that will be used in this chapter.

1. Assume that $v_{0,n}(x)$ and $f_n(x,t)$ are two sequences of nonnegative, bounded functions which have uniformly bounded norms in $L^1(\Omega)$ and $L^1(Q_T)$ (for every $T > 0$), respectively. Then, if one considers the solutions v_n of problems

$$\begin{cases} (v_n)_t - \Delta v_n = f_n(x,t) & \text{in } Q \\ v_n(x,t) = 0 & \text{on } \partial\Omega\times(0,\infty), \\ v_n(x,0) = v_{0,n}(x) & \text{in } \Omega\,, \end{cases}$$

the following estimates hold:

$$\|v_n\|_{L^{r_1}(0,\tau;W^{1,q_1}(\Omega))} \leq C(\tau), \quad \text{for every } (r_1, q_1) \text{such that}$$

$$1 \leq q_1 < \frac{N}{N-1}, \ 1 \leq r_1 \leq 2 \text{ and } \frac{N}{q_1} + \frac{2}{r_1} > N+1\,; \tag{2.8}$$

$$\|v_n\|_{C^0(0,\tau;L^1(\Omega))} \leq C(\tau)\,; \tag{2.9}$$

$$\|T_k v_n\|_{L^2(0,\tau;W_0^{1,2}(\Omega))} \leq C(\tau)k\,, \quad \text{for every } k > 0; \tag{2.10}$$

$$\iint_{Q_\tau} \frac{|\nabla v_n|^2}{(v_n+1)^\alpha} \leq C(\tau, \alpha)\,, \quad \text{for every } \alpha > 1. \tag{2.11}$$

2. Moreover, if f_n converges to some μ in the weak sense of measures in Q_τ, for every $\tau > 0$, and $v_{0,n}$ converges to v_0 in $L^1(\Omega)$, then for every $\tau > 0$

$$v_n \to v \quad \text{in } L^{r_1}(0,\tau; W_0^{1,q_1}(\Omega)), \text{ for every } (r_1, q_1) \text{as in (2.8)}\,, \tag{2.12}$$

where v is the unique solution of

$$\begin{cases} (v)_t - \Delta v = \mu & \text{in } Q \\ v(x,t) = 0 & \text{on } \partial\Omega \times (0,\infty), \\ v(x,0) = v_0(x) & \text{in } \Omega\,, \end{cases}$$

in the sense that

$$\iint_Q (-v\,\phi_t + \nabla v \cdot \nabla \phi)\,dx\,dt - \int_\Omega v_0(x)\,\phi(x,0)\,dx = \langle \mu\,,\,\phi \rangle$$

for every $\phi(x,t) \in C^1(\overline{Q})$ with compact support in $\Omega \times [0,\infty)$. Moreover, if $\mu = \mu(x,t)$ is a function in $L^1_{\text{loc}}(\overline{Q})$, then

$$v \in C^0([0,\infty); L^1(\Omega))\,.$$

Finally, if $f_n \to \mu$ strongly in $L^1(Q_T)$, for $T > 0$, then the truncation verifies,

$$T_k v_n \to T_k v \quad \text{strongly in } L^2(0,T; W_0^{1,2}(\Omega)), \text{ for every } k. \tag{2.13}$$

The same convergence holds if $f_n \rightharpoonup \mu$ in the weak-$*$ convergence of measures, if μ is concentrated on a set of null parabolic capacity, see Sect. 4 below.

The details of these estimates can be found, for instance, in the following works, [34, 42, 62, 100] or in [131]. In this references even a more general framework is also considered.

2.2 Existence of Solution with Higher Regularity

In this section we deal with the problem

$$\begin{cases} u_t - \Delta u = |\nabla u|^2 + f(x,t) & \text{in } Q \\ u(x,t) = 0 & \text{on } \partial\Omega \times (0,\infty), \\ u(x,0) = u_0(x) & \text{in } \Omega, \end{cases} \tag{2.14}$$

where u_0 and f are positive measurable functions.

Assume that

1. f is a positive function such that

$$f(x,t) \in L^r_{\text{loc}}([0,\infty); L^q(\Omega)), \qquad \text{with } q,\ r > 1, \quad \frac{N}{q} + \frac{2}{r} < 2.$$

2. $v_0(x) = e^{u_0} - 1 \in L^2(\Omega)$.

We perform the change of variable $v = e^u - 1$; then problem (2.1) becomes

$$\begin{cases} v_t - \Delta v = f(x,t)(v+1) & \text{in } Q \\ v(x,t) = 0 & \text{on } \partial\Omega \times (0,\infty) \\ v(x,0) = v_0(x) = e^{u_0} - 1. \end{cases} \tag{2.15}$$

Then by using classical arguments as in [97] or [109], there exists a unique solution to (2.15),

$$v \in \mathcal{C}([0,\infty); L^2(\Omega)) \cap L^2_{\text{loc}}([0,\infty); W_0^{1,2}(\Omega)).$$

Using the linearity of the problem the result can be easily adapted to the case where v_0 only belongs to $L^1(\Omega)$, obtaining $v \in \mathcal{C}([0,\infty); L^1(\Omega)) \cap L^2_{\text{loc}}([0,\infty); W_0^{1,2}(\Omega))$. Actually v and ∇v are Hölder continuous (see the classical theory, again in [97]).

Putting $u = \log(v+1)$, then $u \in L^2(0,T; W_0^{1,2}(\Omega))$ and u satisfies problem (2.14). The inverse is also true in the sense that if u is a solution to problem (2.14) with $e^{u_0(x)} - 1 \in L^2(\Omega)$ and $e^u - 1 \in L^2((0,T), W_0^{1,2}(\Omega))$, then if we set $v = e^u - 1$ we obtain that v solves problem (2.15).

2.3 Regularity of General Solutions

We will assume in problem (2.14), u_0 and f are positive functions such that $u_0 \in L^1(\Omega)$ and $f \in L^1_{loc}(\overline{Q})$. The first summability result is the following.

Proposition 2.5 *Assume that* $u \in \mathcal{C}([0,\infty); L^1(\Omega)) \cap L^2_{loc}([0,\infty); W_0^{1,2}(\Omega))$ *is a solution of problem (2.14), where* $f \in L^1_{loc}(\overline{Q})$ *is such that* $f(x,t) \geq 0$ *a.e. in* Q. *Then*

$$\int_\Omega e^{u(x,\tau)} d(x)\, dx < \infty \qquad \textit{for every } \tau > 0, \tag{2.16}$$

where $d(x) = dist(x, \partial\Omega)$.

Proof Let $\varepsilon > 0$, we consider $v_\varepsilon = H_\varepsilon(u)$, where $H_\varepsilon(s) = e^{\frac{s}{1+\varepsilon s}} - 1$, then $v_\varepsilon \in L^\infty(Q) \cap L^2_{loc}([0,\infty); W_0^{1,2}(\Omega))$. We claim that v_ε satisfies the inequality

$$(v_\varepsilon)_t - \Delta v_\varepsilon \geq 0$$

in the sense of distributions. Indeed, we consider positive and smooth approximations in L^1, ϕ_n, f_n and $u_{0,n}$ of $|\nabla u|^2$, f and u_0, respectively, and we consider the approximate problems,

$$\begin{cases} (u_n)_t - \Delta u_n = \phi_n + f_n & \text{in } Q \\ u_n(x,t) = 0 & \text{on } \partial\Omega \times (0,\infty), \\ u_n(x,0) = u_{0,n}(x) & \text{in } \Omega\,, \end{cases}$$

and define

$$v_{n,\varepsilon} = H_\varepsilon(u_n).$$

Then, for every positive $\xi(x,t) \in \mathcal{C}_0^\infty(Q)$,

$$\begin{aligned} &- \iint_Q v_{n,\varepsilon}\, \xi_t \, dx\, dt + \iint_Q \nabla v_{n,\varepsilon} \cdot \nabla \xi \, dx\, dt \\ &\quad = \iint_Q (\phi_n + f_n) H'_\varepsilon(u_n)\, \xi \, dx\, dt - \iint_Q |\nabla u_n|^2 H''_\varepsilon(u_n)\, \xi \, dx\, dt\,. \end{aligned} \tag{2.17}$$

Step 1 Past to the limit in n for fixed ε.

By the theory for parabolic equations with data in L^1, the sequence $\{u_n\}$ satisfies the properties stated in the previous section, and in particular, using convergence (2.12), we can pass to the limit in n in every term of (2.17). As far as the last integral is concerned, one has

$$\iint_Q |\nabla u_n|^2 H_\varepsilon''(u_n)\,\xi\,dx\,dt = \iint_Q |\nabla T_k u_n|^2 H_\varepsilon''(u_n)\,\xi\,dx\,dt + \iint_{\{u_n>k\}} |\nabla u_n|^2 H_\varepsilon''(u_n)\,\xi\,dx\,dt\,.$$

The first integral of the r.h.s. passes to the limit by convergence (2.13), while the second one is small if k is large, uniformly in n, since

$$|H_\epsilon''(s)| \le \frac{c(\epsilon)}{(1+\epsilon s)^3} \qquad \text{for all positive } s\,,$$

and thus, using estimate (2.11),

$$\iint_{\{u_n>k\}} |\nabla u_n|^2\,|H_\epsilon''(u_n)|\,\xi\,dx\,dt \le \frac{c}{(1+\epsilon k)} \iint_{Q\cap\operatorname{supp}\xi} \frac{|\nabla u_n|^2}{(1+\epsilon u_n)^2}\,dx\,dt \le \frac{c}{(1+\epsilon k)}\,.$$

Therefore, since $H_\varepsilon'(u) - H_\varepsilon''(u) \ge 0$, one has

$$-\iint_Q v_\varepsilon\,\xi_t + \iint_Q \nabla v_\varepsilon\cdot\nabla\xi\,dx\,dt = \iint_Q (H_\varepsilon'(u) - H_\varepsilon''(u))\,|\nabla u|^2\,\xi\,dx\,dt + \iint_Q f\,H_\varepsilon'(u)\,\xi\,dx\,dt \ge 0\,.$$

Step 2 Since $u \in C([0,\infty); L^1(\Omega))$, therefore $v_\varepsilon \in C([0,\infty); L^p(\Omega))$ for every $p<\infty$.

Since $u \in L^1(\Omega)$, in particular $e^{u(x,t)} < \infty$ a.e. in Q. For $t_0 > 0$, let w be the solution of problem

$$\begin{cases} w_t - \Delta w = 0 & \text{in } \Omega\times(t_0,\infty)\\ w(x,t) = 0 & \text{on } \partial\Omega\times(t_0,\infty),\\ w(x,t_0) = v_\varepsilon(x,t_0). \end{cases} \tag{2.18}$$

Using Lemma 2 of the Y. Martel paper [108], we find that

$$c_1(t)||v_\varepsilon(\cdot,t_0)d(\cdot)||_{L^1}d(x) \le w(x,t) \le c_2(t)||v_\varepsilon(\cdot,t_0)d(\cdot)||_{L^1}d(x) \quad \text{for all} \quad t>t_0,$$

for some positive functions $c_1(t)$, $c_2(t)$. Since v_ε is a supersolution to problem (2.18), we conclude that $w \le v_\varepsilon$ in $\Omega\times(t_0,\infty)$. Therefore

$$c_1(t)||v_\varepsilon(\cdot,t_0)d(\cdot)||_{L^1}d(x) \le v_\varepsilon(x,t) \le e^{u(x,t)} < \infty \quad \text{for a.e. } (x,t)\in\Omega\times(t_0,\infty)\,.$$

We fix $(x, t) \in \Omega \times (t_0, \infty)$, such that $u(x, t) < \infty$. Then using Fatou's lemma we obtain

$$\int_\Omega e^{u(x,t_0)}\, d(x)\, dx < \infty.$$

Using the fact that $t_0 > 0$ is arbitrary, we conclude that (2.16) holds. □

As a consequence we obtain the following result.

Theorem 2.6 *Under the same hypotheses as in the previous propositions, for all $\tau > 0$ we have*

$$\iint_{Q_\tau} |\nabla u|^2 e^{\delta u}\, dx\, dt < \infty, \qquad \textit{for all } \delta < 1, \tag{2.19}$$

$$\iint_{Q_\tau} f\, e^u\, dx\, dt < \infty, \tag{2.20}$$

$$\iint_{Q_\tau} e^{\frac{u}{1+\varepsilon u}} |\nabla u|^2 \left(1 - \frac{1}{(1+\varepsilon u)^2}\right) dx\, dt \le C(\tau) \quad \textit{uniformly in } \varepsilon, \tag{2.21}$$

$$\int_\Omega e^{u_0(x)}\, dx < \infty. \tag{2.22}$$

and finally

$$e^u \in L^\infty(0, \tau; L^1(\Omega)), \tag{2.23}$$

Proof Let us consider an open set $\tilde{\Omega} \supset\supset \Omega$. For $\tau > 0$, consider the solution $\phi(x, t)$ of problem

$$\begin{cases} -\phi_t - \Delta\phi = 0 & \text{in } \tilde{\Omega} \times (0, \tau + 1) \\ \phi(x, t) = 0 & \text{on } \partial\tilde{\Omega} \times (0, \tau + 1), \\ \phi(x, \tau + 1) = \tilde{d}(x), \end{cases} \tag{2.24}$$

where

$$\tilde{d}(x) = \begin{cases} \operatorname{dist}(x, \partial\Omega) & \text{if } x \in \Omega, \\ 0 & \text{if } x \in \tilde{\Omega} \setminus \Omega. \end{cases}$$

Then by the weak parabolic Harnack inequality (see [110])

$$\phi(x, t) \ge c(\tau) > 0, \qquad \text{for a.e. } (x, t) \in \Omega \times (0, \tau). \tag{2.25}$$

Let us define

$$k_{\delta,\varepsilon}(s) = e^{\frac{\delta s}{1+\varepsilon s}}, \quad \Psi_{\delta,\varepsilon}(s) = \int_0^s k_{\delta,\varepsilon}(\sigma)d\sigma \le \frac{1}{\delta}e^{\delta s}.$$

We use $\phi(x,t)\,(k_{\delta,\varepsilon}(u(x,t)) - 1)$ as test function in (2.14), and we integrate in $Q_{\tau+1}$, obtaining

$$\begin{aligned}
&\int_\Omega \Psi_{\delta,\varepsilon}(u(x,\tau+1))\,d(x)\,dx - \int_\Omega u(x,\tau+1)\,d(x)\,dx \\
&\quad - \int_\Omega \Psi_{\delta,\varepsilon}(u(x,0))\,\phi(x,0)\,dx + \int_\Omega u(x,0)\,\phi(x,0)\,dx \\
&\quad + \iint_{Q_{\tau+1}} k'_{\delta,\varepsilon}(u)\,|\nabla u|^2\,\phi\,dx\,dt = \iint_{Q_{\tau+1}} k_{\delta,\varepsilon}(u)\,|\nabla u|^2\,\phi\,dx\,dt \\
&\quad - \iint_{Q_{\tau+1}} |\nabla u|^2\,\phi\,dx\,dt + \iint_{Q_{\tau+1}} f\,k_{\delta,\varepsilon}(u)\,\phi\,dx\,dt \\
&\quad - \iint_{Q_{\tau+1}} f\,\phi\,dx\,dt\,.
\end{aligned} \tag{2.26}$$

The first integral in (2.26) is bounded by (2.16), therefore, using (2.25), it follows that

$$\begin{aligned}
&\iint_{Q_\tau} e^{\frac{\delta u}{1+\varepsilon u}}\left(1 - \frac{\delta}{(1+\varepsilon u)^2}\right)|\nabla u|^2\,dx\,dt + \iint_{Q_\tau} e^{\frac{\delta u}{1+\varepsilon u}} f\,dx\,dt \\
&\quad + \int_\Omega \Psi_{\delta,\varepsilon}(u_0(x))\,dx = \iint_{Q_\tau} \left(k_{\delta,\varepsilon}(u) - k'_{\delta,\varepsilon}(u)\right)|\nabla u|^2\,dx\,dt \\
&\quad + \iint_{Q_\tau} f\,k_{\delta,\varepsilon}(u)\,dx\,dt + \int_\Omega \Psi_{\delta,\varepsilon}(u_0(x))\,dx \le c(\tau)\,.
\end{aligned}$$

Then, taking $\delta < 1$ and passing to the limit as $\varepsilon \to 0$, we obtain (2.19). Similarly, taking $\delta = 1$, we obtain (2.20), (2.21) and (2.22). Finally, let $\omega(x,t)$ be the solution of

$$\begin{cases}
-\omega_t - \Delta\omega = 0 & \text{in } Q_\tau \\
\omega(x,t) = 0 & \text{on } \partial\Omega \times (0,\tau), \\
\omega(x,\tau) \equiv 1\,.
\end{cases} \tag{2.27}$$

Then $0 \le \omega(x,t) \le 1$ for every $(x,t) \in Q_\tau$. By using in Eq. (2.14) $k_{1,\varepsilon}(u)\,\omega$ as a test function we find that,

$$\int_\Omega \Psi_{1,\varepsilon}(u(x,\tau))\,dx$$
$$\le \iint_{Q_\tau} \left(k_{1,\varepsilon}(u) - k'_{1,\varepsilon}(u)\right)|\nabla u|^2\,dx\,dt + \iint_{Q_\tau} f\,k_{1,\varepsilon}(u)\,dx\,dt + \int_\Omega \Psi_{1,\varepsilon}(u_0(x))\,dx\,. \tag{2.28}$$

Since the right-hand side of (2.28) is bounded by the previous estimates, (2.23) follows. □

Remark 2.7 If we consider the following approximating problem

$$\begin{cases} (u_n)_t - \Delta u_n = \dfrac{|\nabla u|^2}{1+\frac{1}{n}|\nabla u|^2} + T_n(f(x,t)) & \text{in } Q \\ u_n(x,t) = 0 & \text{on } \partial\Omega \times (0,\infty), \\ u_n(x,0) = T_n(u_0(x)), \end{cases}$$

then we can prove using the previous regularity results that $u_n \uparrow u$ and $u_n \to u$ strongly in $L^2(0,\tau; W_0^{1,2}(\Omega))$ for all $\tau > 0$.

2.3.1 Optimality of the Hypotheses on f: Nonexistence Result

To see that the condition on f is in some sense optimal, we will assume that $0 \in \Omega$ and that $f(x,t) = f(x) = \dfrac{\lambda}{|x|^2}$. Note that $f(x) \in L^q(\Omega)$ for every $q < N/2$, therefore we are in a limit case of (1). Hence we have the following nonexistence result.

Theorem 2.8 *Assume that $N \ge 3$, and that $\lambda > \Lambda_N = \left(\dfrac{N-2}{2}\right)^2$, the optimal Hardy constant defined by*

$$\Lambda_N \equiv \inf_{\{\phi \in W_0^{1,2}(\Omega);\ \phi \ne 0\}} \frac{\displaystyle\int_\Omega |\nabla\phi|^2 dx}{\displaystyle\int_\Omega \phi^2\,|x|^{-2}\,dx}.$$

Then, for any initial datum $u_0 \ge 0$ and for any $T > 0$, problem

$$\begin{cases} u_t - \Delta u = |\nabla u|^2 + \dfrac{\lambda}{|x|^2} & \text{in } Q_T \\ u(x,t) = 0 & \text{on } \partial\Omega \times (0,T), \\ u(x,0) = u_0(x) & \text{in } \Omega\,, \end{cases} \tag{2.29}$$

has no solution.

Proof The proof uses the same arguments as in [44] and [3] (see also [126]); for the sake of completeness we include here the proof. We argue by contradiction. Assume that u is a solution to problem (2.29) with $f(x,t) = \dfrac{\lambda}{|x|^2}$, $\lambda > \Lambda_N$. Let $\phi \in C_0^\infty(\Omega)$, by taking ϕ^2 as a test function in (2.1) we obtain that

$$\begin{aligned}\int_\Omega u(x,t_2)\,\phi^2\,dx - \int_\Omega u(x,t_1)\,\phi^2 dx + 2\iint_{Q_{t_1,t_2}} \phi\,\nabla\phi\cdot\nabla u\,dx\,dt \\ = \iint_{Q_{t_1,t_2}} \phi^2\,|\nabla u|^2\,dx\,dt + \lambda\iint_{Q_{t_1,t_2}} \frac{\phi^2}{|x|^2}\,dx\,dt,\end{aligned}$$

where we have set $Q_{t_1,t_2} = \Omega\times(t_1,t_2)$. Hence

$$\begin{aligned}-\int_\Omega u(x,t_2)\,\phi^2\,dx &\le 2\iint_{Q_{t_1,t_2}} \phi\,\nabla\phi\cdot\nabla u\,dx\,dt - \iint_{Q_{t_1,t_2}} \phi^2\,|\nabla u|^2\,dx\,dt \\ &-\lambda\iint_{Q_{t_1,t_2}} \frac{\phi^2}{|x|^2}\,dx\,dt = -\iint_{Q_{t_1,t_2}} |\nabla\phi - \phi\,\nabla u|^2\,dx\,dt \\ &+\iint_{Q_{t_1,t_2}} |\nabla\phi|^2\,dx\,dt - \lambda\iint_{Q_{t_1,t_2}} \frac{\phi^2}{|x|^2}\,dx\,dt \\ &\le (t_2-t_1)\Big[\int_\Omega |\nabla\phi|^2\,dx - \lambda\int_\Omega \frac{\phi^2}{|x|^2}\,dx\Big].\end{aligned}$$

By the regularity result of Theorem 2.6, we know that $u(\cdot,t)\in L^p(\Omega)$ for all $t\in(0,T)$ and for all $p<\infty$; therefore we obtain that

$$\int_\Omega |\nabla\phi|^2\,dx - \lambda\int_\Omega \frac{\phi^2}{|x|^2}\,dx \ge -\frac{1}{t_2-t_1}\Big(\int_\Omega u^{\frac{N}{2}}(x,t_2)\,dx\Big)^{\frac{2}{N}}\Big(\int_\Omega |\phi|^{2^*}\,dx\Big)^{\frac{2}{2^*}}.$$

By density, this implies that

$$I(\Omega) \equiv \inf_{\phi\in W_0^{1,2}(\Omega)\setminus\{0\}} \frac{\displaystyle\int_\Omega |\nabla\phi|^2\,dx - \lambda\int_\Omega \frac{\phi^2}{|x|^2}\,dx}{\Big(\displaystyle\int_\Omega |\phi|^{2^*}\,dx\Big)^{\frac{2}{2^*}}} \ge -\frac{1}{t_2-t_1}\Big(\int_\Omega u^{\frac{N}{2}}(x,t_2)\,dx\Big)^{\frac{2}{N}} > -\infty.$$

Since $\lambda > \left(\frac{N-2}{2}\right)^2$, taking the sequence $\phi_n(x) = T_n(|x|^{-\frac{N-2}{2}})\eta(x)$, where $\eta(x)$ is a cut-off function with compact support in Ω which is 1 in a neighborhood of the origin, one can check that $I(\Omega) = -\infty$. Hence we reach a contradiction. □

Corollary 2.9

(1) *If* $f(x,t) \geq \dfrac{C(t)}{|x|^{2+\varepsilon}}$ *in a neighborhood of the origin, where* $C(t)$ *is a positive function such that* $C(t) \geq a > 0$ *in* $(t_1, t_2) \subset (0, T)$, *then problem* (2.29) *has no solution.*

(2) *Since the argument used in the proof is local, then under the same hypothesis on* f *we can prove that problem* (2.29) *has no local positive solution.*

Proof It suffices to observe that in this case, for any $\lambda > \Lambda_N$, one has $f(x,t) \geq \dfrac{\lambda}{|x|^2}$ in a small ball centered at the origin. □

2.4 Existence of Weaker Solution and Nonuniqueness

As a consequence of the regularity obtained in Sects. (2.2) and (2.3), we will prove some connection between the nonlinear problem (2.1) and some linear problems with measure data. This will give one of the main results in this work: a non-uniqueness result for problem (2.14) which is the counterpart in the evolutionary problem of the non-uniqueness result obtained for the stationary problem in the previous chapter.

2.4.1 Nonuniqueness: Existence of Weaker Solutions

In this first part of the section we will show a strong connection between solutions of problem (2.14) and solutions of a linear problem with measure datum.

The theory of parabolic equations in divergence form with measure data has been strongly developed in the last 40 years. See for instance [23, 29, 32, 34, 42, 61, 99, 100, 131] and references therein. Various definitions of solution have been introduced in order to obtain uniqueness results and it is well known that the uniqueness is still an open problem for general nonlinear operators.

However in the case of problem (2.31) below, the situation is easier, as far as uniqueness is concerned, because we are considering the heat operator and then the different kind of solutions coincide.

The first result we will prove is an existence and uniqueness theorem for problem (2.15) with an additional term which is a finite Radon measure.

We say that a measurable function defined on $\Omega \times (0, \infty)$ verifies the *Aronson-Serrin condition* if and only if

$$f \in L^r_{\text{loc}}([0,\infty); L^q(\Omega)) \text{ with } r, q > 1 \text{ such that } \frac{N}{2q} + \frac{1}{r} < 1. \tag{2.30}$$

Theorem 2.10 *Let f be a function verifying the Aronson-Serrin condition. Let μ be a Radon measure on Q, which is finite on Q_T for every $T > 0$. Then problem*

$$\begin{cases} v_t - \Delta v = f(x,t)\, v + \mu & \text{in } Q \\ v = 0 & \text{on } \partial\Omega \times (0,\infty), \\ v(x,0) = \phi(x) \in L^1(\Omega), \end{cases} \tag{2.31}$$

has a unique distributional solution such that

$$\begin{cases} (i) \quad v \in L^{r_1}_{\mathrm{loc}}([0,\infty); W_0^{1,q_1}(\Omega)) \text{ for every } r_1, q_1 \geq 1 \text{ such that } \dfrac{N}{q_1} + \dfrac{2}{r_1} > N+1\,; \\ (ii) \quad v \in L^{\infty}_{\mathrm{loc}}([0,\infty); L^1(\Omega))\,, \text{ for every } k > 0; \\ (iii) \quad T_k v \in L^2_{\mathrm{loc}}([0,\infty); W_0^{1,2}(\Omega))\,, \text{ for every } k > 0; \\ (iv) \quad f\, v \in L^1_{\mathrm{loc}}(\overline{Q})\,. \end{cases} \tag{2.32}$$

Proof If v satisfies (2.32) (i) and (ii), then, using the Gagliardo-Nirenberg inequality, $v \in L^{\rho}_{\mathrm{loc}}([0,\infty); L^{\sigma}(\Omega))$, for all ρ and σ satisfying

$$\rho, \sigma \geq 1\,, \qquad \frac{N}{\sigma} + \frac{2}{\rho} > N\,. \tag{2.33}$$

Consider $g_n \in L^\infty(Q)$, such that $\{g_n\}$ is bounded in $L^1(Q_T)$ for every $T > 0$ and moreover, as $n \to \infty$,

$$g_n \rightharpoonup \mu \quad \text{weakly in the measures sense in } Q_T, \text{ for every } T > 0.$$

Consider $\phi_n \in L^\infty(\Omega)$, $\phi_n \to \phi$ in $L^1(\Omega)$. We solve

$$\begin{cases} (v_n)_t - \Delta v_n = f\, v_n + g_n & \text{in } Q \\ v_n = 0 & \text{on } \partial\Omega \times (0,\infty), \\ v_n(x,0) = \phi_n(x). \end{cases}$$

Claim For every $T > 0$ there exists a constant $C(T) > 0$ such that

$$||v_n||_{L^{r'}(0,T;L^{q'}(\Omega))} \leq C(T) \qquad \frac{1}{r} + \frac{1}{r'} = 1,\ \frac{1}{q} + \frac{1}{q'} = 1.$$

where r, q are as in (2.30).

If the claim holds then $f\, v_n$ is uniformly bounded in $L^1(Q_T)$ for every $T > 0$ and we can conclude in a standard way (see for instance [34] and [42]). Hence it is sufficient to prove the claim above.

We argue by contradiction; assume that, up to a subsequence,

$$||v_n||_{L^{r'}(0,T;L^{q'}(\Omega))} \to \infty.$$

Normalizing the sequence, i.e., putting $w_n = \dfrac{v_n}{||v_n||_{L^{r'}(0,T;L^{q'}(\Omega))}}$, then $||w_n||_{L^{r'}(0,T;L^{q'}(\Omega))} = 1$ and for each $n \in \mathbb{N}$, w_n satisfies problem

$$\begin{cases} (w_n)_t - \Delta w_n = f(x,t)\, w_n + \dfrac{g_n}{||v_n||_{L^{r'}(0,T;L^{q'}(\Omega))}} & \text{in } Q_T \\ w_n = 0 & \text{on } \partial\Omega \times (0,T), \\ w_n(x,0) = \dfrac{\phi_n(x)}{||v_n||_{L^{r'}(0,T;L^{q'}(\Omega))}}. \end{cases}$$

The right-hand side in Eq. (2.4.1) is uniformly bounded in $L^1(Q_T)$, hence by using the results (2.8)–(2.11) in Sect. 2 we find that $\{w_n\}$ is bounded in $L^\infty(0,T;L^1(\Omega))$ and in $L^{r_1}(0,T;W_0^{1,q_1}(\Omega))$, for all (r_1,q_1) as in (2.32) *(i)*. Therefore by Sobolev's embedding, $\{w_n\}$ is bounded in $L^\rho(0,T;L^\sigma(\Omega))$, for all (ρ,σ) as in (2.33). Hence there exists w such that $w_n \rightharpoonup w$ weakly in $L^{r_1}(0,T;W_0^{1,q_1}(\Omega))$ for all (r_1,q_1) as in (2.32) *(i)*. Moreover, w verifies

$$\begin{cases} w_t - \Delta w = f(x,t)\, w & \text{in } Q_T \\ \quad w \in L^\infty(0,T;L^1(\Omega)) \cap L^{r_1}(0,T;W_0^{1,q_1}(\Omega)) \text{ for all } (r_1,q_1) \text{as in (2.32)}, \\ w(x,0) = 0, \end{cases} \tag{2.34}$$

because

$$\frac{g_n}{||v_n||_{L^{r'}(0,T;L^{q'}(\Omega))}} \to 0 \text{ in } L^1(Q_T) \text{ and } \frac{\phi_n(x)}{||v_n||_{L^{r'}(0,T;L^{q'}(\Omega))}} \to 0 \text{ in } L^1(\Omega), \text{ as } n \to \infty.$$

We will proceed in two steps.

Step 1) $w_n \to w$ *strongly in* $L^{r'}(0,T;L^{q'}(\Omega))$, *therefore* $||w||_{L^{r'}(0,T;L^{q'}(\Omega))} = 1$.

Indeed, by using the compact embedding $W_0^{1,q_1}(\Omega) \hookrightarrow L^s(\Omega)$ if $s < q_1^*$, the continuous embedding $L^s(\Omega) \subset W^{-1,q_1'}(\Omega) + L^1(\Omega)$ and the fact that

$$||w_n||_{L^{r_1}(0,T;W_0^{1,q_1}(\Omega))} \le C \text{ and } ||(w_n)_t||_{L^{r_1'}(0,T;W^{-1,q_1'}(\Omega))+L^1(Q_T)} \le C,$$

using Aubin's compactness results (see for instance [137]), we conclude that $\{w_n\}$ is relatively compact in $L^{r_1}(0,T;L^s(\Omega))$ for all $s < q_1^*$. Therefore, $\{w_n\}$ is relatively compact in $L^\rho(0,T;L^\sigma(\Omega))$ for all (ρ,σ) as in (2.33). Therefore we only have to show that one can take $(\rho, \sigma) = (r', q')$ in (2.33). Indeed, the condition

$$\frac{N}{q'} + \frac{2}{r'} > N$$

is equivalent to the assumption (2.30). This argument proves the Step 1.

Step 2) *Problem* (2.34) *admits only the trivial solution.*

Since uniqueness is trivial in the space $L^2(0, T;\ W_0^{1,2}(\Omega))$, we only have to show that every solution w of (2.34) belongs to this space. This is done by a bootstrap method. Indeed, using Hölder's inequality and the regularity of f we find that $f w^{m_1} \in L^1(Q_T)$, for every m_1 such that

$$\frac{N}{m_1 q'} + \frac{2}{m_1 r'} > N,$$

and since $\frac{1}{q'} + \frac{2}{Nr'} > 1$, we can chose $1 < m_1 < \frac{1}{q'} + \frac{2}{Nr'}$. Therefore, using $\frac{w^{m_1-1}}{1+\varepsilon w^{m_1-1}}$ as a test function in (2.34) and passing to the limit as $\varepsilon \to 0$, we obtain, for every $\tau \in (0, T)$,

$$\frac{1}{m_1} \int_\Omega w^{m_1}(x, \tau)\, dx + (m_1 - 1) \iint_{Q_\tau} w^{m_1-2}\, |\nabla w|^2\, dx\, dt = \iint_{Q_\tau} f\, w^{m_1}\, dx\, dt$$
$$\leq \iint_{Q_\tau} f\, w^{m_1}\, dx\, dt = C(T),\ \forall \tau \in [0, T]\,,$$

Hence, setting

$$v = w^{\frac{m_1}{2}}$$

the last estimate implies

$$v \in L^2((0, T);\ W_0^{1,2}(\Omega)) \cap L^\infty((0, T);\ L^2(\Omega))\,,$$

which by Gagliardo-Nirenberg inequality gives

$$v \in L^\delta((0, T);\ L^\gamma(\Omega)) \text{ with } 2 \leq \gamma \leq 2^*,\ \delta \leq 2 \text{ and } \frac{2}{\delta} + \frac{N}{\gamma} = \frac{N}{2}.$$

Hence it follows that $w \in L^\beta((0, T);\ L^\alpha(\Omega))$ where

$$\alpha = \frac{m_1 \gamma}{2},\ \beta = \frac{\delta m_1}{2},\ m_1 \leq \alpha \leq \frac{2^*}{2} m_1,\ m_1 \leq \beta \text{ and } \frac{m_1}{\beta} + \frac{N m_1}{2\alpha} = \frac{N}{2}. \tag{2.35}$$

This implies that

$$\iint_{Q_T} f w^{m_2}\,dx\,dt < \infty\,, \qquad \text{where } m_2 = m_1\,(\frac{1}{q'} + \frac{2}{Nr'})\,.$$

Iterating the process, if we consider the sequence defined by

$$m_{k+1} = \rho\, m_k\,, \quad \text{with } \rho = \frac{1}{q'} + \frac{2}{Nr'} > 1\,,$$

then

$$\iint_{Q_T} f\, w^{m_k}\,dx\,dt < \infty\,.$$

Thus

$$\iint_{Q_T} w^{m_k-2}\,|\nabla w|^2\,dx\,dt < C(k) \text{ and } \sup_{\tau\in(0,T)} \int_\Omega w^{m_k}(x,t)dx < C(k).$$

As $m_k \to \infty$ and since $T_k(w) \in L^2(0,T; W_0^{1,2}(\Omega))$, then for $k > 1$, it follows that

$$\iint_{Q_T} |\nabla w|^2\,dx\,dt \le \iint_{Q_T} |\nabla T_k(w)|^2\,dx\,dt + \iint_{Q_T} w^{m_k-2}\,|\nabla w|^2\,dx\,dt < C(k).$$

Thus $w \in L^2(0,T; W_0^{1,2}(\Omega)) \cap L^\infty(0,T; L^2(\Omega))$ and then the uniqueness result follows.

Finally, Step 1) and Step 2) give a contradiction, and then the claim is proved.

□

The previous problem (2.31) with measure datum appears in a natural way when we perform the change of unknown function as before. Theorems 2.13 and 2.15 below will show that there exists a one-to-one correspondence between the solutions of problem (2.14) and (2.31), where μ is an arbitrary "singular" measure. To clarify the meaning of "singular" measure we have to use a notion of *parabolic capacity* introduced by Pierre in [128] and by Droniou, Porretta and Prignet in [62].

For $T > 0$, we define the Hilbert space $\mathbf{W}$ by setting

$$\mathbf{W} = \mathbf{W}_T = \{u \in L^2(0,T; W_0^{1,2}(\Omega)),\; u_t \in L^2(0,T; W^{-1,2}(\Omega))\},$$

equipped with the norm defined by

$$||u||^2_{\mathbf{W}_T} = \iint_{Q_T} |\nabla u|^2\,dx\,dt + \int_0^T ||u_t||^2_{W^{-1,2}}dt\,.$$

Definition 2.11 If $U \subset Q_T$ is an open set, we define

$$\mathrm{cap}_{1,2}(U) = \inf\{\|u\|_{\mathbf{W}} : u \in \mathbf{W},\ u \geq \chi_U \text{ almost everywhere in } \quad Q_T\}$$

(we will use the convention that $\inf \emptyset = +\infty$), then for any borelian subset $B \subset Q_T$ the definition is extended by setting:

$$\mathrm{cap}_{1,2}(B) = \inf\left\{\mathrm{cap}_{1,2}(U),\ U \text{ open subset of} Q_T, B \subset U\right\}.$$

We refer to [62] for the main properties of this capacity. We observe that, if $B \subset Q_T \subset Q_{\tilde{T}}$, then the capacity of B is the same in Q_T and in $Q_{\tilde{T}}$, therefore we will not specify the value of T when speaking of a Borel set compactly contained in $\overline{Q}$.

We recall that, given a Radon measure μ on Q and a Borel set $E \subset Q$, then μ is said to be concentrated on E if $\mu(B) = \mu(B \cap E)$ for every Borel set B.

Definition 2.12 Let the space dimension N be at least 2. Let μ be a positive Radon measure in Q. We will say that μ is singular if it is concentrated on a subset $E \subset Q$ such that

$$\mathrm{cap}_{1,2}(E \cap Q_\tau) = 0, \text{ for every } \tau > 0.$$

As examples of singular measures, one can consider:

(i) a space-time Dirac delta $\mu = \delta_{(x_0,t_0)}$ defined by $\langle \mu, \varphi \rangle = \varphi(x_0, t_0)$ for every $\varphi(x,t) \in \mathcal{C}_c(Q)$;
(ii) a Dirac delta in space $\mu = \mu(x) = \delta_{x_0}$ defined by $\langle \mu, \varphi \rangle = \int_0^\infty \varphi(x_0, t)\, dt$;
(iii) more generally, a measure μ concentrated on the set $E \times (0, +\infty)$, where $E \subset \Omega$ has zero "elliptic" 2-capacity;
(iv) a measure μ concentrated on a set of the form $E \times \{t_0\}$, where $E \subset \Omega$ has zero Lebesgue measure.

The main result of this chapter is the following multiplicity result.

Theorem 2.13 *Let μ_s be a positive, singular Radon measure such that $\mu_s\big|_{Q_T}$ is bounded for every $T > 0$. Assume that $f(x,t)$ is a positive function such that $f \in L^r_{\mathrm{loc}}([0,\infty); L^q(\Omega))$, where r and q satisfy the Aronson-Serrin hypothesis* (2.30), *and that the initial datum u_0 satisfies $v_0 = e^{u_0} - 1 \in L^1(\Omega)$. Consider v, the unique solution of problem*

$$\begin{cases} v_t - \Delta v = f(x,t)\,(v+1) + \mu_s \text{ in } \mathcal{D}'(Q) \\ \qquad v \in L^\infty_{\mathrm{loc}}([0,\infty); L^1(\Omega)) \cap L^\rho_{\mathrm{loc}}([0,\infty); W_0^{1,\sigma}(\Omega)) \\ \qquad\qquad \text{where } \sigma, \rho > 1 \text{ verify } \dfrac{N}{\sigma} + \dfrac{2}{\rho} > N+1 \\ v(x,0) = v_0(x), \qquad f\, v \in L^1_{\mathrm{loc}}(\overline{Q}). \end{cases} \tag{2.36}$$

We set $u = \log(v+1)$, *then* $u \in L^2_{\mathrm{loc}}([0,\infty); W^{1,2}_0(\Omega)) \cap C([0,\infty); L^1(\Omega))$ *and is a weak solution of*

$$\begin{cases} u_t - \Delta u = |\nabla u|^2 + f(x,t) \text{ in } \mathcal{D}'(Q) \\ u(x,0) = u_0(x) \equiv \log(v_0(x)+1)\,. \end{cases} \tag{2.37}$$

Proof Let $h_n(x,t) \in L^\infty(Q)$ be a sequence of bounded nonnegative functions such that

- $\|h_n\|_{L^1(Q_T)} \le C(T)$ for every $T > 0$, and
- $h_n \rightharpoonup \mu_s$ weakly in the measures sense in Q_T, for every $T > 0$.

Consider now the unique solution v_n to problem

$$\begin{cases} (v_n)_t - \Delta v_n = T_n(f\,(v+1)) + h_n \qquad \text{in } Q \\ v_n \in L^2_{\mathrm{loc}}([0,\infty); W^{1,2}_0(\Omega)) \\ v_n(x,0) = T_n(v_0(x))\,. \end{cases} \tag{2.38}$$

Notice that $(v_n)_t \in L^2_{\mathrm{loc}}(\overline{Q})$ (see for instance [63]), and that, for every $T > 0$, $v_n \to v$ in $L^\rho(0,T; W^{1,\sigma}_0(\Omega))$ for all ρ and σ as in (2.36). We set $u_n = \log(v_n+1)$, then by a direct computation one can check that

$$(u_n)_t - \Delta u_n = |\nabla u_n|^2 + \frac{T_n(f\,(v+1))}{v_n+1} + \frac{h_n}{v_n+1} \text{ in } \mathcal{D}'(Q).$$

Notice that by using the definition of v_n we conclude easily that, for every $T > 0$,

$$\frac{T_n(f(v+1))}{v_n+1} \to f(x,t) \text{ in } L^1(Q_T) \text{ and } u_n \to u \text{ in } L^1(Q_T). \tag{2.39}$$

We claim that

$$\frac{h_n}{v_n+1} \to 0 \text{ in } \mathcal{D}'(Q). \tag{2.40}$$

To prove the claim let $\phi(x,t)$ be a function in $C^\infty_0(Q)$; we want to prove that

$$\lim_{n\to\infty} \iint_{Q_T} \phi \frac{h_n}{v_n+1}\,dx = 0\,.$$

We assume that supp $\phi \subset Q_T$, and we use the assumption on μ_s: let $A \subset Q_T$ be such that $\mathrm{cap}_{1,2}(A) = 0$ and $\mu_s \llcorner Q_T$ is concentrated on A.

Then for all $\varepsilon > 0$ there exists an open set $U_\varepsilon \subset Q_T$ such that $A \subset U_\varepsilon$ and $\mathrm{cap}_{1,2}(U_\varepsilon) \le \varepsilon/2$. Then, we can find a function $\psi_\varepsilon \in \mathbf{W}_T$ such that $\psi_\varepsilon \ge \chi_{U_\varepsilon}$ and

$||\psi_\varepsilon||_{\mathbf{W}_T} \le \varepsilon$. Let us define the real function

$$m(s) = \frac{2|s|}{|s|+1}.$$

Then one has

$$m(\psi_\varepsilon) \le 2, \qquad m(\psi_\varepsilon) \ge \chi_{U_\varepsilon} \text{ and}$$

$$\iint_{Q_T} |\nabla m(\psi_\varepsilon)|^2 \, dx \, dt = \iint_{Q_T} |m'(\psi_\varepsilon)|^2 |\nabla \psi_\varepsilon|^2 \, dx \, dt \le 4\,\varepsilon^2.$$

Using a Picone-type inequality in Theorem 1.1, we obtain that

$$4\,\varepsilon^2 \ge \int_\Omega |\nabla m(\psi_\varepsilon)|^2 \, dx \ge \int_\Omega \frac{-\Delta(v_n+1)}{v_n+1} m^2(\psi_\varepsilon) \, dx \ge \int_\Omega \frac{h_n}{v_n+1} m^2(\psi_\varepsilon) \, dx$$
$$- \int_\Omega \frac{(v_n)_t}{v_n+1} m^2(\psi_\varepsilon) \, dx.$$

By integration in t, we get

$$\begin{aligned}\iint_{U_\varepsilon} \frac{h_n}{v_n+1} \, dx \, dt &\le 4\,\varepsilon^2\, T + \textstyle\int_\Omega \log(v_n(x,T)+1)\, m^2(\psi_\varepsilon(x,T)) \, dx \\ &\quad + 2 \iint_{Q_T} \log(v_n+1)\, m(\psi_\varepsilon)\, m'(\psi_\varepsilon)\, (\psi_\varepsilon)_t \, dx \, dt \\ &= 4\,\varepsilon^2\, T + I_1 + I_2 .\end{aligned} \tag{2.41}$$

(1) Estimate of I_1 Since $|m(s)| \le 2$, then using Hölder's inequality we obtain that

$$I_1 \le C\Big(\int_\Omega \log^2(v_n(x,T)+1) \, dx \Big)^{\frac{1}{2}} \Big(\int_\Omega m^4(\psi_\varepsilon(x,T)) \, dx \Big)^{\frac{1}{2}} \le C\Big(\int_\Omega m^2(\psi_\varepsilon(x,T)) \, dx \Big)^{\frac{1}{2}}$$

where in the last estimate we have used the inequality $\log(s+1) \le s^{\frac{1}{2}} + c$ and the bound

$$\max_{t\in[0,T]} \int_\Omega v_n(x,t) \, dx \le C(T).$$

Since $m(s) \le 2\,|s|$, it follows that

$$I_1 \le C\Big(\int_\Omega |\psi_\varepsilon(x,T)|^2 \, dx \Big)^{\frac{1}{2}} \le \max_{t\in[0,T]} \Big(\int_\Omega \psi_\varepsilon^2(x,t) \, dx \Big)^{\frac{1}{2}} \le C\, ||\psi_\varepsilon||_{\mathbf{W}_T} \le C\,\varepsilon, \tag{2.42}$$

by the fact that $\mathbf{W}_T \subset \mathcal{C}([0,T]; L^2(\Omega))$ with a continuous inclusion.

(2) Estimate of I_2 Using $\dfrac{m^2(\psi_\varepsilon)}{v_n+1}$ as a test function in (2.38), by a direct computation we obtain

$$\begin{aligned}
&\int_\Omega \log(v_n(x,T)+1)\, m^2(\psi_\varepsilon(x,T))\,dx - \int_\Omega \log(T_n(v_0)+1)\, m^2(\psi_\varepsilon(x,0))\,dx\\
&\quad -2\iint_{Q_T} \log(v_n+1)\, m(\psi_\varepsilon)\, m'(\psi_\varepsilon)\,(\psi_\varepsilon)_t\,dx\,dt\\
&\quad +2\iint_{Q_T} m(\psi_\varepsilon)\, m'(\psi_\varepsilon)\,\nabla\psi_\varepsilon\,\frac{\nabla v_n}{v_n+1}\,dx\,dt\\
&\quad -\iint_{Q_T} m^2(\psi_\varepsilon)\,\frac{|\nabla v_n|^2}{(v_n+1)^2}\,dx\,dt\\
&\quad =\iint_{Q_T} \frac{m^2(\psi_\varepsilon)}{v_n+1}\Big(T_n(f(v+1))+h_n(x,t)\Big)\,dx\,dt \ge 0\,.
\end{aligned}$$

Thus, recalling (2.42) and (2.11) which holds for v_n, we get

$$\begin{aligned}
&2\iint_{Q_T} \log(v_n+1)\, m(\psi_\varepsilon)\, m'(\psi_\varepsilon)\,(\psi_\varepsilon)_t\,dx\,dt\\
&\quad \le I_1 + 2\iint_{Q_T} m(\psi_\varepsilon)\, m'(\psi_\varepsilon)\,\nabla\psi_\varepsilon\frac{\nabla v_n}{v_n+1}\,dx\,dt \le C\,\varepsilon + 8\iint_{Q_T}|\nabla\psi_\varepsilon|\frac{|\nabla v_n|}{v_n+1}\,dx\,dt\\
&\quad \le C\,\varepsilon + 8\Big(\iint_{Q_T}|\nabla\psi_\varepsilon|^2\,dx\,dt\Big)^{\frac12}\Big(\iint_{Q_T}\frac{|\nabla v_n|^2}{(v_n+1)^2}\,dx\,dt\Big)^{\frac12} \le C\,\varepsilon\,.
\end{aligned} \tag{2.43}$$

Hence by (2.41) we conclude that

$$\iint_{U_\varepsilon} \frac{h_n}{v_n+1}\,dx\,dt \le C(\varepsilon+\varepsilon^2)\,. \tag{2.44}$$

Now, by (2.43),

$$\begin{aligned}
&\Big|\iint_{Q_T} \phi\frac{h_n}{v_n+1}\,dx\,dt\Big|\\
&\quad \le ||\phi||_\infty \iint_{U_\varepsilon}\frac{h_n}{v_n+1}\,dx\,dt + \iint_{Q_T\setminus U_\varepsilon}|\phi|\,h_n\,dx\,dt \le C||\phi||_\infty\,(\varepsilon+\varepsilon^2)\\
&\quad + \iint_{Q_T\setminus U_\varepsilon}|\phi|\,h_n\,dx\,dt\,.
\end{aligned}$$

Since $h_n \to \mu_s$ in $\mathcal{M}_0(Q_T)$ and μ_s is concentrated on $A \subset U_\varepsilon$, we conclude that

$$\iint_{\Omega\setminus U_\varepsilon} |\phi|\, h_n\, dx\, dt \to 0 \quad \text{as } n \to \infty.$$

Since ε is arbitrary we get the desired result, hence the claim (2.40) follows.

Let $\phi \in C_0^\infty(Q_T)$, then we have

$$\begin{aligned}&\iint_{Q_T} ((u_n)_t - \Delta u_n)\, \phi\, dx\, dt\\ &= \iint_{Q_T} \frac{T_n(f(v+1))}{v_n+1}\phi\, dx\, dt + \iint_{Q_T} |\nabla u_n|^2 \phi\, dx\, dt + \iint_{Q_T} \frac{h_n \phi}{v_n+1}\, dx\, dt.\end{aligned}$$

Hence using (2.39) and (2.40) we just have to prove that

$$|\nabla u_n|^2 \to |\nabla u|^2 \text{ in } L^1(Q_T)$$

which means that

$$\frac{|\nabla v_n|^2}{(v_n+1)^2} \to \frac{|\nabla v|^2}{(v+1)^2} \text{ in } L^1(Q_T).$$

Since the sequence $\{\frac{|\nabla v_n|^2}{(v_n+1)^2}\}$ converges a.e. in Q_T to $\frac{|\nabla v|^2}{(v+1)^2}$, then by Vitali's theorem we only have to prove that it is equi-integrable. Let $E \subset Q_T$ be a measurable set. Then, for every $\delta \in (0,1)$ and $k > 0$,

$$\begin{aligned}\iint_E \frac{|\nabla v_n|^2}{(v_n+1)^2}\, dx\, dt &= \iint_{E\cap\{v_n\le k\}} \frac{|\nabla v_n|^2}{(v_n+1)^2}\, dx\, dt + \iint_{E\cap\{v_n> k\}} \frac{|\nabla v_n|^2}{(v_n+1)^2}\, dx\, dt\\ &\le \iint_E |\nabla T_k(v_n)|^2\, dx\, dt + \frac{1}{(k+1)^{1-\delta}} \iint_{Q_T} \frac{|\nabla v_n|^2}{(v_n+1)^{1+\delta}}\, dx\, dt.\end{aligned}$$

By (2.11), the last integral is uniformly bounded with respect to n, therefore the corresponding term can be made small by choosing k large enough.

Moreover, since μ_s is singular and

$$T_n(f(v+1)) \to f(v+1) \text{ in } L^1(Q_T),$$

by (2.13) we have

$$T_k(v_n) \to T_k(v) \text{ strongly on } L^2(0,T; W_0^{1,2}(\Omega)) \text{ for any } k > 0,$$

therefore the integral

$$\int_E |\nabla T_k(v_n)|^2 \, dx\,dt \text{ is uniformly small if meas } (E) \text{ is small enough.}$$

The equi-integrability of $|\nabla u_n|^2$ follows immediately, and the proof is completed. Hence we conclude that

$$u_t - \Delta u = |\nabla u|^2 + f(x,t) \text{ in } \mathcal{D}'(Q).$$

Since $|\nabla u|^2 + f \in L^1(\Omega \times (0,T))$, then using classical result about the regularity and uniqueness of entropy solution we obtain that $u \in \mathcal{C}([0,\infty); L^1(\Omega))$ and the result follows. □

Remark 2.14

1. Notice that if we consider $x_0 \in \Omega$ and $0 < t_0 < T$ and the problem

$$v_t - \Delta u = \delta_{x_0,t_0}, \quad v(x,t) = 0 \text{ on } \partial\Omega \times (0,T), \quad v(x,0) = 0,$$

then it is easy to check that $t \to ||v(t)||_1$, has a jump in $t = t_0$. However, defining $u = \log(1+v)$, u belongs to $\mathcal{C}([0,T]; L^1(\Omega))$. The mechanism of this behavior is as follows:

(i) u solves the equation $u_t - \Delta u = |\nabla u|^2$ in the sense of distributions;
(ii) the regularity theory for L^1 data provides the continuity.

2. In general we can prove that if v is a solution to problem

$$v_t - \Delta v = \mu \text{ in } Q_T, \quad v(x,0) = v_0(x) \in L^1(\Omega),$$

where μ is a positive Radon measure, then $\sup_{t\in[0,T]} \int_\Omega v(x,t)\,dx \le C(\mu(Q_T), T)$. Indeed, consider ω, the solution to problem (2.27), it is clear that $\omega \le 1$, hence ω is globally defined and therefore using ω as a test function in (3.35), it follows that

$$\int_\Omega v(x,\tau)\,dx \le \int_\Omega v_0(x)\,\omega(x,0)\,dx + c(T)\mu(Q_T).$$

Hence the result follows by taking the maximum for $\tau \in [0,T]$.

2.4.2 The Converse Result: Characterization of the Measure

We will prove that every solution of problem (2.1), $u \in \mathcal{C}([0,\infty); L^1(\Omega)) \cap L^2_{loc}([0,\infty); W_0^{1,2}(\Omega))$, by the Hopf-Cole transformation determines in a unique way a *capacity-singular measure* in Q. Therefore we will obtain a one-to-one correspondence between solutions to problem (2.1) and singular measures with respect the capacity in Q.

Theorem 2.15 *Let* $u \in C([0,\infty); L^1(\Omega)) \cap L^2_{\text{loc}}([0,\infty); W_0^{1,2}(\Omega))$ *be a solution to problem (2.14), where* $f(x,t)$ *is a positive function such that* $f \in L^r_{\text{loc}}([0,\infty); L^q(\Omega))$*, where* r *and* q *satisfy the Aronson-Serrin hypothesis (2.30). Consider* $v = e^u - 1$*, then* $v \in L^1_{\text{loc}}(\overline{Q})$*, and there exists a bounded positive measure* μ *in* Q_T *for every* $T > 0$*, such that*

1. μ *is concentrated on a set* A *and* $\text{cap}_{1,2}(A \cap Q_T) = 0$ *for all* $T > 0$*, that is* μ *is a singular measure with respect to* $\text{cap}_{1,2}$*-capacity.*
2. v *is a distributional solution of*

$$v_t - \Delta v = f(x,t)\,(v+1) + \mu \qquad \text{in } Q\,. \tag{2.45}$$

Moreover μ *can be characterized as a weak limit in the space of bounded Radon measures, as follows:*

$$\mu = \lim_{\epsilon \to 0} |\nabla u|^2 e^{\frac{u}{1+\varepsilon u}} \left(1 - \frac{1}{(1+\varepsilon u)^2}\right) \quad \text{in } Q_T, \text{for every } T > 0. \tag{2.46}$$

Proof We set $v = e^u - 1$, then by the regularity results of Theorem 2.6, we obtain that $v \in L^1_{\text{loc}}(\overline{Q})$ and

$$\iint_{Q_\tau} f(x,t)\,(v+1)\,dx\,dt + \iint_{Q_\tau} |\nabla u|^2\, e^{\frac{u}{1+\varepsilon u}} \left(1 - \frac{1}{(1+\varepsilon u)^2}\right) dx\,dt \le C(\tau)\,. \tag{2.47}$$

Therefore, there exists a positive Radon measure μ in Q such that for all $\tau > 0$

$$|\nabla u|^2 e^{\frac{u}{1+\varepsilon u}} \left(1 - \frac{1}{(1+\varepsilon u)^2}\right) \rightharpoonup \mu \quad \text{in the weak measure sense in} Q_\tau.$$

Notice that for every $k > 0$

$$\iint_{Q_\tau \cap \{u \le k\}} |\nabla u|^2\, e^{\frac{u}{1+\varepsilon u}} \left(1 - \frac{1}{(1+\varepsilon u)^2}\right) dx\,dt \to 0 \text{ as } \varepsilon \to 0.$$

That means that μ is concentrated in the set $A \equiv Q \setminus \bigcup_{k>0} \{u \le k\}$. We now define

$$v_\varepsilon(x,t) = \int_0^{u(x,t)} e^{\frac{s}{1+\varepsilon s}}\, ds \in L^2_{\text{loc}}([0,\infty); W_0^{1,2}(\Omega)).$$

By making an approximation as in the first part of the proof of Proposition 2.5, it is easy to check that v_ε solves

$$(v_\varepsilon)_t - \Delta v_\varepsilon = e^{\frac{u}{1+\varepsilon u}}|\nabla u|^2(1 - \frac{1}{(1+\varepsilon u)^2}) + f(x,t)e^{\frac{u}{1+\varepsilon u}} \tag{2.48}$$

in the sense of distributions.

By (2.47) and the monotone convergence theorem we get easily that the last term converges in $L^1(Q_\tau)$ for all $\tau > 0$, while the remaining one converges to μ. Since $v_\varepsilon \to v$ in $L^1(Q_\tau)$ for all $\tau > 0$, we obtain that v solves Eq. (2.45) in the sense of distributions, therefore μ is uniquely determined.

Finally to prove that $\text{cap}_{1,2}(A \cap Q_T) = 0$ and then μ is a singular measure in the sense of Definition 2.12 we use the Lemma 2.17 in [62] as follows.

Consider $A_T = A \cap Q_T$, and remember that $u \in \mathcal{C}([0,T]; L^1(\Omega)) \cap L^2([0,T]; W_0^{1,2}(\Omega))$ solves problem

$$\begin{cases} u_t - \Delta u = g(x,t) \equiv |\nabla u|^2 + f(x,t) & \text{in } Q_T \\ u(x,t) = 0 & \text{on } \partial\Omega \times (0,T), \\ u(x,0) = u_0(x) & \text{in } \Omega, \end{cases}$$

Let $\tau \le T$, using $T_k(u)$, defined in (1.13), as a test function in the above problem it follows that

$$\int_\Omega \Theta_k(u(x,\tau))\,dx + \iint_{Q_\tau} |\nabla T_k(u)|^2\,dx\,dt = \iint_{Q_\tau} g(x,t)T_k(u)\,dx\,dt + \int_\Omega \Theta_k(u_0(x))\,dx$$

where

$$\Theta_k(s) = \int_0^s T_k(\sigma)d\sigma = \begin{cases} \frac{1}{2}s^2 & \text{if } |s| \le k, \\ ks - \frac{1}{2}k^2 & \text{if } |s| \ge k. \end{cases}$$

Thus

$$\int_\Omega \Theta_k(u(x,\tau))\,dx + \iint_{Q_\tau} |\nabla T_k(u)|^2\,dx\,dt \le k(||g||_{L^1(Q_T)} + ||u_0||_{L^1(\Omega)}).$$

Since $\Theta_k(s) \ge \dfrac{1}{2}T_k^2(s)$, we conclude that

$$||T_k(u)||^2_{L^\infty((0,T);L^2(\Omega))} + ||T_k(u)||^2_{L^2((0,T);W_0^{1,2}(\Omega))} \le C(T)k.$$

Consider $w_k = \dfrac{T_k(u)}{k}$, it is clear that

$$w_k \in \mathbf{X} \equiv L^\infty((0,T); L^2(\Omega)) \cap L^2((0,T); W_0^{1,2}(\Omega)) \text{ and } ||w_k||_X^2 \le \frac{C(T)}{k}.$$

Hence $||w_k||_{\mathbf{X}}^2 \to 0$ as $k \to \infty$.

Using an approximation argument and by Kato type inequality, see for instance [123], there results that

$$(w_k)_t - \Delta w_k \ge 0.$$

Therefore by using Lemma 2.17 in [62], we obtain $z_k \in \mathbf{W}$ such that

$$z_k \ge w_k \text{ and } ||z_k||_{\mathbf{W}} \le ||w_k||_{\mathbf{X}}.$$

It is clear that $z_k \ge 1$ on A_T. Hence

$$\text{cap}_{1,2}(A_T) \le ||z_k||_{\mathbb{W}} \le ||w_k||_X \le (\frac{C(T)}{k})^{\frac{1}{2}}.$$

Letting $k \to \infty$ it follows that $\text{cap}_{1,2}(A_T) = 0$ and then the result follows. □

Corollary 2.16 *There exist a unique solution to problem* (2.14) *in the class*

$$\mathcal{X} = \{u \in L^1_{\text{loc}}(Q) \,:\, e^{\frac{u}{2}} - 1 \in L^2_{\text{loc}}([0,\infty); W_0^{1,2}(\Omega))\}.$$

Proof It is sufficient to observe that, setting $v = e^u - 1$, then by Theorem 2.15, v solves (2.45). Using (2.46) we get $\mu = 0$.

We claim that

$$\int_\Omega v(x,\tau)\phi dx \to \int_\Omega (e^{u_0(x)} - 1)\phi dx \text{ as } \tau \to 0 \text{ for all } \phi \in \mathcal{C}^1(\overline{\Omega}),\ \ \phi|_{\partial\Omega} = 0.$$

From the regularity result of Theorem (2.6) we know that $e^u \in L^\infty(0,\tau; L^1(\Omega))$. Let $\phi \in \mathcal{C}^1(\overline{\Omega})$ be such that $\phi|_{\partial\Omega} = 0$, since $e^{\frac{u}{2}} - 1 \in L^2_{\text{loc}}([0,\infty); W_0^{1,2}(\Omega))$, then using Theorem (2.6), and by an approximation argument, we can use $e^u\phi$ as a test function in (2.14). Hence it follows that

$$\int_\Omega e^{u(x,\tau)}\phi\, dx + \int_0^\tau \int_\Omega e^u \nabla u \nabla \phi\, dx\, dt = \int_0^\tau \int_\Omega e^u f \phi\, dx\, dt + \int_\Omega e^{u_0(x)}\phi\, dx.$$

Since $f e^u \in L^1(Q_{\tau_1})$ where $\tau_1 > 0$, then

$$\lim_{\tau\to 0} \int_0^\tau \int_\Omega e^u f \phi dx\, dt = 0.$$

Moreover we have

$$\int_0^\tau \int_\Omega e^u |\nabla u||\nabla \phi| dx\, dt \le \frac{1}{2}\int_0^\tau \int_\Omega e^u |\nabla u|^2\, dx\, dt + \int_0^\tau \int_\Omega e^u |\phi|^2\, dx\, dt \to 0 \text{ as } \tau \to 0.$$

Putting together the previous estimates we conclude that

$$\int_\Omega v(x,\tau)\phi dx = \int_\Omega (e^{u(x,\tau)} - 1)\, \phi\, dx \to \int_\Omega (e^{u_0(x)} - 1)\phi\, dx \text{ as } \tau \to 0$$

and then the claim follows. Hence $v \in L^2_{loc}([0,\infty); W_0^{1,2}(\Omega))$ solves

$$v_t - \Delta v = f(x,t)\,(v+1) \qquad \text{in } Q\,.$$

with

$$\int_\Omega v(x,\tau)\phi\, dx \to \int_\Omega (e^{u_0(x)} - 1)\phi\, dx \text{ as } \tau \to 0.$$

The linear classical theory gives the uniqueness. □

2.4.3 Nonuniqueness Induced by Singular Perturbations of the Initial Data

We prove in this section nonuniqueness for problem (2.14) by perturbing the initial data in the associated linear problem with a suitable singular measure. For sake of simplicity, we limit ourselves to the case where $f(x,t) \equiv 0$. In what follows, we will denote by $|E|$ the usual Lebesgue measure $\mathbb{R}^N$. The main result in this direction is the following.

Theorem 2.17 *Let ν_s be a bounded positive singular measure in Ω, concentrated on a subset $E \subset\subset \Omega$ such that $|E| = 0$. Let v be the unique solution of problem*

$$\begin{cases} v_t - \Delta v = 0 \textit{ in } \mathcal{D}'(Q) \\ v(x,t) = 0 \textit{ on } \partial\Omega \times (0,\infty) \\ v(x,0) = \nu_s\,. \end{cases} \tag{2.49}$$

We set $u = \log(v+1)$, then $u \in L^2_{loc}([0,\infty); W_0^{1,2}(\Omega))$ and verifies

$$\begin{cases} u_t - \Delta u = |\nabla u|^2 \textit{ in } \mathcal{D}'(Q) \\ u(x,0) = 0. \end{cases} \tag{2.50}$$

Proof Let $h_n \in L^\infty(\Omega)$ be a sequence of nonnegative functions such that $||h_n||_{L^1(\Omega)} \le C$ and $h_n \rightharpoonup \nu_s$ weakly in the measure sense, namely

$$\lim_{n\to\infty} \int_\Omega h_n(x)\phi(x)dx \to \langle \nu_s, \phi\rangle \text{ for all } \phi \in C_c(\Omega).$$

Consider now v_n the unique solution to problem

$$\begin{cases} (v_n)_t - \Delta v_n = 0 \text{ in } Q \\ \qquad v_n \in L^2_{\rm loc}([0,\infty); W_0^{1,2}(\Omega)) \\ \quad v_n(x,0) = h_n(x)\,. \end{cases} \tag{2.51}$$

Notice that $v_n \to v$ strongly in $L^r(0,T; W_0^{1,q}(\Omega))$ for all r and q satisfying $\dfrac{N}{q}+\dfrac{2}{r} > N+1$ and

$$\int_\Omega v_n(x,t)\phi(x)dx \to \int_\Omega h_n(x)\phi(x)dx \text{ as } t\to 0,\ \text{ for all } \phi \in C(\overline{\Omega}).$$

As in the proof of Theorem 2.13, we can prove that $|\nabla u_n|^2 \to |\nabla u|^2$ strongly in $L^1(Q_T)$ for all $T > 0$, the only difference being that in this case the strong convergence of the truncates is proved in [30].

Moreover to finish we have just to show that $\log(1+v_n(.,t)) \to 0$ strongly in $L^1(\Omega)$ as $t \to 0$ and $n \to \infty$.

To prove this last affirmation, take $H(v_n)$, where $H(s) = 1 - \dfrac{1}{(1+s)^\alpha}$, $0 < \alpha << 1$, as a test function in (2.51), then

$$\int_\Omega \overline{H}(v_n(x,\tau))\,dx + \alpha \iint_{Q_\tau} \frac{|\nabla v_n|^2}{(1+v_n)^{1+\alpha}}\,dx\,dt = \int_\Omega \overline{H}(h_n(x))\,dx$$

where $\overline{H}(s) = \int_0^s H(\sigma)d\sigma = s - \frac{1}{1-\alpha}((1+s)^{1-\alpha}-1)$. Hence $\int_\Omega v_n(x,t)\,dx \le C$ where C is positive constant independent of n and t. As a consequence we obtain that $\log(1+v_n(.,t))$ is bounded in $L^p(\Omega)$ for all $p < \infty$ uniformly in n and t.

By the strong convergence of $T_k v_n$, then for small $\varepsilon > 0$ we get the existence of $n(\varepsilon)$ and $\tau(\varepsilon) > 0$ such that for $n \ge n(\varepsilon)$ and $t \le \tau(\varepsilon)$, we have

$$\iint_{Q_t} \frac{|\nabla v_n|^2}{(1+v_n)^2}\,dx\,ds \le \varepsilon. \tag{2.52}$$

Since ν_s is concentrated on a set $E \subset\subset \Omega$ with $|E| = 0$, then for $\varepsilon \in (0,1)$ there exists an open set U_ε such that $E \subset U_\varepsilon \subset \Omega$ and $|U_\varepsilon| \le \varepsilon/2$.

Without loss of generality we can assume that $\operatorname{supp} h_n \subset U_\varepsilon$ for $n \ge n(\varepsilon)$.

Let $\phi_\varepsilon \in \mathcal{C}_0^\infty(\mathbb{R}^N)$ be such that $0 \le \phi_\varepsilon \le 1$, $\phi_\varepsilon = 1$ in U_ε, $\operatorname{supp}\phi_\varepsilon \subset O_\varepsilon$ and $|O_\varepsilon| \le 2\varepsilon$.

Consider w_ε, the solution to problem

$$\begin{cases} w_{\varepsilon t} - \Delta w_\varepsilon = 0 \text{ in } Q, \\ \quad w_\varepsilon(x,t) = 0 \text{ on } \partial\Omega \times (0,\infty), \\ \quad w_\varepsilon(x,0) = \phi_\varepsilon(x). \end{cases}$$

It is clear that $0 \le w_\varepsilon \le 1$,

$$w_\varepsilon \to 0 \text{ strongly in } L^2(0,\infty);\, W_0^{1,2}(\Omega)),$$

$$w_\varepsilon \to 0 \text{ strongly in } \mathcal{C}([0,\infty);\, L^2(\Omega)),$$

(and $\dfrac{dw_\varepsilon}{dt} \to 0$ strongly in $L^2(0,\infty);\, W^{-1,2}(\Omega))$).

For $t \le \tau \equiv \tau(\varepsilon)$, we set $\widetilde{w}_\varepsilon(x,t) = w(x,\tau - t)$, using $\dfrac{\widetilde{w}_\varepsilon}{1+v_n}$ as a test function in (2.51), it follows that

$$\int_\Omega \log(1+v_n(x,\tau))\, \widetilde{w}_\varepsilon(x,\tau)\, dx - \iint_{Q_\tau} \frac{|\nabla v_n|^2}{(1+v_n)^2} \widetilde{w}_\varepsilon\, dx\, ds = \int_\Omega \log(1+h_n) \widetilde{w}_\varepsilon(x,0)\, dx.$$

Using (2.52) and the properties of $\widetilde{w}_\varepsilon$, we get

$$\int_{U_\varepsilon} \log(1+v_n(x,\tau))\, dx \le \varepsilon + \int_\Omega \log(1+h_n)\, \widetilde{w}_\varepsilon(x,0)\, dx \le \varepsilon + \int_\Omega \log(1+h_n)\, dx$$

It is clear that we can obtain the same estimate for any $t \le \tau(\varepsilon)$. Since $\operatorname{supp} h_n \subset U_\varepsilon$, then

$$\int_\Omega \log(1+h_n)\, dx = \int_{U_\varepsilon} \log(1+h_n)\, dx \le C\Big(\varepsilon + \int_{U_\varepsilon} h_n^{1/2}\, dx\Big) \le C(\varepsilon + \varepsilon^{1/2}) \le C\,\varepsilon^{1/2},$$

Hence we conclude that

$$\int_{U_\varepsilon} \log(1+v_n(x,t))\, dx \le C\,\varepsilon^{1/2} \text{ for } n \ge n(\varepsilon) \text{ and } t \le \tau(\varepsilon). \tag{2.53}$$

We now deal with the complement integral $\int_{\Omega\setminus U_\varepsilon} \log(1+v_n(x,t))\, dx$.

Let $\psi_\varepsilon \in C^\infty(\mathbb{R}^N)$ be such that $0 \leq \psi_\varepsilon \leq 1$, $\psi_\varepsilon = 0$ in N where N is an open set such that $E \subset\subset N \subset\subset U_\varepsilon$ and $\psi_\varepsilon \equiv 1$ in $\Omega \backslash U_\varepsilon$.

As above, let z_ε, the solution to problem

$$\begin{cases} (z_\varepsilon)_t - \Delta z_\varepsilon = 0 \text{ in } Q, \\ z_\varepsilon(x,t) = 0 \text{ on } \partial\Omega \times (0,\infty), \\ z_\varepsilon(x,0) = \psi_\varepsilon(x). \end{cases}$$

It is not difficult to see that $0 \leq z_\varepsilon \leq 1$. For $t \leq \tau \equiv \tau(\varepsilon)$, we consider $\widetilde{z}_\varepsilon(x,t) = z(x, \tau - t)$, using $\dfrac{\widetilde{z}_\varepsilon}{1+v_n}$ as a test function in (2.51), and proceeding as above, we get the existence of $\tau(\varepsilon)$ and $n(\varepsilon)$ such that for $n \geq n(\varepsilon)$ and $t \leq \tau(\varepsilon)$, we have

$$\int_\Omega \log(1 + v_n(x,t))\, dx \leq C\varepsilon^{1/2}$$

and then we get the desired result.

Hence, as a conclusion we obtain that u solves (2.50). □

Remarks 2.18 The previous theorem can also be shown to be true under the presence of an additional initial data $v_0 \in L^1(\Omega)$ and a term $f(x,t)$ in the right-hand side. Therefore, putting together this and the result of Theorem 2.13, the following general multiplicity result can be proved.

Assume that μ_s is a positive Radon measure in Q, singular with respect to the parabolic capacity $\text{cap}_{1,2}$, and $\nu_s \in M(\Omega)$ is a positive Radon measure in Ω, singular with respect to the classical Lebesgue measure, and let v be the unique positive solution to problem

$$\begin{cases} v_t - \Delta v = f(x,t)\,(v+1) + \mu_s \text{ in } \mathcal{D}'(Q) \\ v(x,0) = v_0(x) + \nu_s\,, \end{cases}$$

where $f \in L^r_{\text{loc}}([0,\infty); L^q(\Omega))$, with r and q satisfy the Aronson-Serrin hypothesis (2.30), and $v_0 \in L^1(\Omega)$. If we set $u = \log(1+v)$, then u solves

$$\begin{cases} u_t - \Delta u = |\nabla u|^2 + f(x,t) \text{ in } \mathcal{D}'(Q), \\ u(x,0) = \log(1 + v_0(x)). \end{cases}$$

2.5 Further Results

As in the stationary case, we can analyze the problem

$$\begin{cases} u_t - \Delta u = \beta(u)|\nabla u|^2 + f(x,t) & \text{in } Q \equiv \Omega \times (0,+\infty) \\ u(x,t) = 0 & \text{on } \partial\Omega \times (0,+\infty), \\ u(x,0) = u_0(x) & \text{in } \Omega, \end{cases} \tag{2.54}$$

where f is a nonnegative function in $L^{\infty}_{\text{loc}}(\overline{Q})$ and

$$\beta : [0,\infty) \longrightarrow [0,\infty)$$

is a continuous nondecreasing function, not identically zero. As in (1.33), we consider

$$\gamma(t) = \int_0^t \beta(s)ds, \qquad \Psi(t) = \int_0^t e^{\gamma(s)}ds,$$

then

$$v(x,t) = \Psi(u(x,t)).$$

Then problem (2.1) becomes

$$\begin{cases} v_t - \Delta v = f(x,t)\, g(v) & \text{in } Q \\ v = 0 & \text{on } \partial\Omega \times (0,\infty) \\ v(x,0) = \Psi(u_0) & \text{in } \Omega, \end{cases} \tag{2.55}$$

where g is defined in (1.35) and verifies the properties proved in Sect. 1.4.

Proposition 2.19 *Assume that g verifies the assumptions above and that f is a bounded function. Let v_0 be a bounded positive function, then there exists a unique positive solution $v \in L^{\infty}_{\text{loc}}(\overline{Q})$ to problem*

$$\begin{cases} v_t - \Delta v = f(x,t)\, g(v) & \text{in } Q \\ v = 0 & \text{on } \partial\Omega \times (0,\infty) \\ v(x,0) = v_0(x) & \text{in } \Omega. \end{cases} \tag{2.56}$$

Therefore problem (2.1) *has at least one positive solution u such that*

$$\Psi(u) \in L^{\infty}_{\text{loc}}(\overline{Q}) \cap L^2_{\text{loc}}([0,\infty); W^{1,2}_0(\Omega))$$

and $u(x,0) = \Psi^{-1}(v_0)$.

Proof The proof is trivial, using a sub/super-solution argument, considering a super-solution of the form $w = w(t)$. By the property of g in (1.36) all solutions of (2.56) with bounded data are bounded in Q_T. Since g is locally Lipschitz, the uniqueness follows directly by using Gronwall's inequality. □

In order to obtain a global solution for unbounded initial data and a measure source term, we will assume the following structural hypotheses on g, which is satisfied by all elementary functions $\beta(u)$:

$$g(s) \le c(1 + sA(\log^* s)), \text{ for every } s > 0, \tag{2.57}$$

where $\log^* s = \max\{\log s, 1\}$, and $A(t) : [0, +\infty) \to [0, +\infty)$ is a continuous, increasing function such that

1. A satisfies the so-called Δ_2 condition, that is,

$$A(2t) \le kA(t) \quad \text{for all } t \ge t_0 \tag{2.58}$$

   for some positive constants k and t_0;
2. A is at most slightly superlinear, in the sense that

$$\int^{+\infty} \frac{ds}{A(s)} = +\infty. \tag{2.59}$$

The following existence result is proved in [59].

Proposition 2.20 *Assume that g verifies* (2.57), (2.58) *and the* (2.59) *condition. If $v_0 \in L^1(\Omega)$, and μ is a positive measure in Q which is bounded in Q_T for every positive T, then there exists a function*

$$v \in L^\infty_{loc}([0, \infty); L^1(\Omega)) \cap L^q_{loc}([0, \infty); W^{1,q}_0(\Omega)) \cap L^\sigma_{loc}(\Omega \times [0, \infty))$$

for every $q < 1 + \frac{1}{N+1}$ and for every $\sigma < 1 + \frac{2}{N}$, such that

(a) For every $\delta < \frac{1}{2}$, $|v|^\delta \in L^2_{loc}([0, \infty); H^1_0(\Omega))$;
(b) For all $k > 0$, $T_k(v) \in L^2_{loc}([0, \infty); H^1_0(\Omega))$,

which is a weak solution to

$$\begin{cases} v_t - \Delta v = f(x, t)\, g(v) + \mu & \text{in } Q \\ v = 0 & \text{on } \partial\Omega \times (0, +\infty) \\ v(x, 0) = v_0(x) & \text{in } \Omega, \end{cases} \tag{2.60}$$

Moreover, if $\mu = 0$ and $v_0 \in L^2(\Omega)$, then

$$v \in C^0([0, \infty); L^2(\Omega)) \cap L^2_{loc}([0, \infty); W^{1,2}_0(\Omega)).$$

Finally, if g satisfies

$$|g(s_1) - g(s_2)| \le C\,(1 + |s_1|^b + |s_2|^b)\,|s_1 - s_2|\,, \qquad 0 < b < \frac{2}{N}\,, \tag{2.61}$$

for every $s_1, s_2 \in \mathbb{R}$*, then the weak solution of* (2.60) *is unique.*

Remark 2.21 The assumptions (2.57), (2.58), (2.59) and (2.61) are satisfied in all the model cases (for instance in the case where $\beta(s)$ is a power, an exponential, or a finite iteration of exponentials, however we do not know whether they hold for every choice of β.

2.5.1 Regularity and Existence of Weaker Solutions

Assume that $f \in L^1_{\rm loc}(\overline{Q})$ is a nonnegative function. Let us consider a distributional solution u of problem (2.1) in the sense of Definition 2.3.

Proposition 2.22 *Assume that* $u(x,t)$ *is a distributional solution of problem (2.1), where* $f \in L^1_{loc}(\overline{Q})$ *is such that* $f(x,t) \ge 0$ *a.e. in* Q*. Then*

$$\int_\Omega \Psi(u(x,t))\,d(x)\,dx < \infty, \qquad \textit{a.e for every } t > 0, \tag{2.62}$$

where Ψ *is defined as in* (1.33).

Proof It suffices to consider the function

$$v_\varepsilon = H_\varepsilon(s) = \int_0^{\frac{s}{1+\varepsilon s}} e^{\gamma(\sigma)}\,d\sigma\,,$$

and to follow the lines of Proposition 2.5, using the inequalities

$$\beta(s)\,H'_\varepsilon(s) - H''_\varepsilon(s) \ge 0\,, \qquad |H''(s)| \le \frac{c(\varepsilon)}{(1+\varepsilon s)^3}\,.$$

□

As a consequence and using the same type of computation as in the proof of Theorem 2.6 we get the following main regularity result.

Theorem 2.23 *Under the same hypotheses as in the previous Propositions, for all* $\tau > 0$ *we have*

$$\iint_{Q_\tau} \beta(u)\,|\nabla u|^2\,e^{\delta\gamma(u)}\,dx\,dt < \infty, \qquad \textit{for all } \delta < 1\,, \tag{2.63}$$

$$\iint_{Q_\tau} f\,e^{\gamma(u)}\,dx\,dt < \infty\,, \tag{2.64}$$

$$\iint_{Q_\tau} \beta(u)\, e^{\frac{\gamma(u)}{1+\varepsilon\gamma(u)}}\, |\nabla u|^2 \left(1 - \frac{1}{(1+\varepsilon\gamma(u))^2}\right) dx\, dt \le C(\tau) \quad \textit{uniformly in } \varepsilon\,, \tag{2.65}$$

$$\int_\Omega \Psi(u_0(x))\, dx < \infty\,, \tag{2.66}$$

and finally

$$\Psi(u(x,t)) \in L^\infty_{\mathrm{loc}}([0,\infty)\,;\, L^1(\Omega))\,. \tag{2.67}$$

Proof It suffices to follow the lines of the proof of Theorem 2.6: first one takes $\phi(x,t)\,\big(k_{\delta,\varepsilon}(u(x,t)) - 1\big)$ as test function in (2.1), where $\phi(x,t)$ is the solution of problem (2.24), and

$$k_{\delta,\varepsilon}(s) = e^{\frac{\delta\gamma(s)}{1+\varepsilon\gamma(s)}}\,, \qquad \delta \le 1\,.$$

Using the inequality (2.62) and passing to the limit as $\varepsilon \to 0$, one obtains (2.63)–(2.66). Then one multiplies by $k_{1,\varepsilon}(u(x,t))\,\omega(x,t)$, with $\omega(x,t)$ satisfying (2.27), to obtain (2.67). □

2.5.2 Existence and Multiplicity Result

The main result of this section is the following.

Theorem 2.24 *Let μ_s be a bounded, positive, singular measure on Q such that $\mu_s(Q_T)$ is bounded for every $T > 0$. Let v be a solution to problem*

$$\begin{cases} v_t - \Delta v = f(x,t)\, g(v) + \mu_s \text{ in } \mathcal{D}'(Q) \\ v \in L^\infty_{\mathrm{loc}}([0,\infty); L^1(\Omega)) \cap L^r_{\mathrm{loc}}([0,\infty); W^{1,q}_0(\Omega)) \\ f(x,t)\, g(v) \in L^1_{\mathrm{loc}}(\overline{Q}) \\ v(x,0) = v_0(x) \in L^1(\Omega)\,, \end{cases} \tag{2.68}$$

for all (r,q) such that

$$q, r \ge 1\,, \qquad \frac{N}{q} + \frac{2}{r} > N+1\,.$$

If we define $u = \Psi^{-1}(v)$, where Ψ is given by (1.33), then u solves

$$\begin{cases} u_t - \Delta u = \beta(u)|\nabla u|^2 + f(x,t) \text{ in } \mathcal{D}'(Q) \\ u \in L^2_{\mathrm{loc}}([0,\infty); W^{1,2}_0(\Omega)) \\ \beta(u)|\nabla u|^2 \in L^1_{\mathrm{loc}}(\overline{Q}) \\ u(x,0) = u_0(x) := \Psi^{-1}(u_0(x))\,. \end{cases} \tag{2.69}$$

The proof is similar to the proof for the case $\beta \equiv 1$ with some technical changes. See [5] for details.

Also the inverse problem hold true, namely we have the following result.

Theorem 2.25 *Let $u \in \mathcal{C}([0,\infty); L^1(\Omega)) \cap L^2_{loc}([0,\infty); W_0^{1,2}(\Omega))$ be a solution to problem (2.1), with $\beta(u)|\nabla u|^2 \in L^1_{loc}(\overline{Q})$ and $f \in L^\infty_{loc}(\overline{Q})$, is a positive function. Let $v = \Psi(u)$, then $v \in L^1_{loc}(Q)$ and there exists a bounded positive Radon measure μ_s, singular with respect to* $\mathrm{cap}_{1,2}$*-capacity, such that v solves*

$$v_t - \Delta v = f(x,t)\, g(v) + \mu_s \quad \text{in } \mathcal{D}'(Q)\,.$$

Moreover μ_s can be characterized as a weak limit in the space of bounded Radon measures, as follows:

$$\mu_s = \lim_{\epsilon \to 0} e^{\frac{\gamma(u)}{1+\varepsilon\gamma(u)}}\, \beta(u)\, |\nabla u|^2 \left(1 - \frac{1}{(1+\varepsilon\gamma(u))^2}\right) \quad \text{in } Q_\tau \text{ for every } \tau > 0.$$

Remark 2.26 Notice that if $\beta \in L^1[0,\infty)$, then necessarily the measure μ_s defined in (2.25) is equivalent to 0. This result follows using the fact that $\gamma(s) \leq \int_0^\infty \beta(\sigma)d\sigma$ and that

$$\lim_{\epsilon \to 0} \iint_{Q_T} \beta(u)|\nabla u|^2 e^{\frac{\gamma(u)}{1+\varepsilon\gamma(u)}} \left(1 - \frac{1}{(1+\varepsilon\gamma(u))^2}\right) \phi\, dx\, dt = 0 \text{ for all } \phi \in \mathcal{C}_0^\infty(Q_T).$$

Moreover if $\beta \in L^1[0,\infty) \cap L^\infty[0,\infty)$, then g is a Lipschitz function, hence problem (2.25) with $\mu_s = 0$ has a unique positive local solution, thus problem (2.1) has a unique local solution. In the elliptic case, the uniqueness result under this condition on β was obtained in [95].

We skip the details and refer to [5].

3 A Kardar-Parisi-Zhang Model in Porous Media and Fast Diffusion Equations

3.1 Introduction

In the paper by Barenblatt et al. [21] is proposed the following equation

$$\gamma h_t = \kappa \frac{\partial^2 (h^2)}{\partial x^2} + \mu \left|\frac{\partial h}{\partial x}\right|^2, \quad h \geq 0,$$

in order to study the ground water flow in a water-absorbing fissured porous rock in one spacial dimension. Here h means the fluid level and γ, κ and μ are parameters characteristic of the medium and the interaction fluid-rock, which are assumed to be constant. The parameter μ can be positive or negative according is the rock non-fissured or fissured respectively.

We will analyze in this chapter the corresponding porous media equation and also the fast diffusion equation, that is, we will drive our attention to the problem

$$\begin{cases} u_t - \Delta u^m = |\nabla u|^q + f(x,t), \quad u \geq 0 & \text{in } \Omega_T \equiv \Omega \times (0,T), \\ u(x,t) = 0 & \text{on } \partial\Omega \times (0,T), \\ u(x,0) = u_0(x), & \text{in } \Omega, \end{cases} \tag{3.1}$$

where $\Omega \subset \mathbb{R}^N$, is a smooth bounded domain, $N \geq 1$, $m > 0$, $1 < q \leq 2$, and $f \geq 0$, $u_0 \geq 0$, are in a suitable class of measurable functions. If $m > 1$, problem (3.1) is a model of growth in a porous medium, see again [21].

We refer to the monograph [145], and the references therein for the basic results about *Porous Media Equations* (PME) and *Fast Diffusion Equation* (FDE) without gradient term. An optimal existence result for the homogeneous Cauchy problem (without source term) can be found in [26].

We will start with the corresponding stationary problem

$$(E) \qquad -\Delta(v^m) = |\nabla v|^q + f,$$

that by the change $v^m = u$ is transformed in the following kind of problems

$$-\Delta u = u^{q\alpha}|\nabla u|^q + \lambda f(x), \quad \alpha = (\frac{1}{m} - 1)$$

Notice that from the porous media equation and fast diffusion equation, we only reach $\alpha \in (-1, \infty)$, however the behavior in the range $(-\infty, -1]$ has theoretical interest in itself.

Therefore, we will study the general problem,

$$\begin{cases} -\Delta u = u^{q\alpha}|\nabla u|^q + \lambda f(x) & \text{in } \Omega \\ u = 0 & \text{on } \partial\Omega, \end{cases} \tag{3.2}$$

where $\Omega \subset \mathbb{R}^N$ is a bounded domain, $f(x) \geq 0$, $\alpha \in (-\infty, \infty)$, $\lambda \geq 0$ and $q \in (1, 2]$.

By the same change of variable the evolution problems,

$$u_t - \Delta u^m = \mu,$$

and (3.1) becomes

$$b(v)_t - \Delta v = \mu \text{ with } b(s) = s^{\frac{1}{m}} \tag{3.3}$$

and

$$b(v)_t - \Delta v = v^{q\alpha}|\nabla v|^q + \lambda f(x), \quad \alpha = (\frac{1}{m} - 1)\ b(s) = s^{\frac{1}{m}}, \tag{3.4}$$

respectively, which we will consider in the last part of this chapter.

Here we are interested in the existence and regularity of solutions to Problems (3.1), (3.2), (3.3) and (3.4), related to source term and to the parameters q and m.

If $q = 2$, some results were obtained in [57] and [79] for bounded data.

It is relevant to note that one of the new features in these kind of models is that we sometimes need to consider singular problems at the boundary and then, in particular, is necessary to give a meaning of how the datum is attained. This fact could be interesting from the mathematical point of view.

The result for the stationary problem are a part of the paper [7] while the results for the evolution problem can be found in [9] and [8]. See also the references therein.

3.2 The Stationary Problem

Consider the problem

$$\begin{cases} -\Delta u = u^{q\alpha}|\nabla u|^q + \lambda f(x), & u \geq 0 \text{ in } \Omega \\ \quad u = 0 & \text{on } \partial\Omega, \end{cases} \tag{3.5}$$

where $\Omega \subset \mathbb{R}^N$ is a bounded domain, $f(x) \geq 0$, $\alpha \in (-\infty, \infty)$ and $q \in (1, 2]$.

We summarize the main results for problem (3.5).

1. If $f \in L^\infty(\Omega)$ and $q(\alpha + 1) \leq 1$, we prove the existence of a solution independently of the size of $||f||_\infty$. On the contrary, in the case $q(\alpha + 1) > 1$, we have to assume λ small.
2. If $q\alpha < -1$ we prove the existence of a positive distributional solution for any L^1 data.
3. The case $-1 \leq q\alpha < 0$ is more regular: by assuming a suitable hypothesis on f we will show that independently of λ, the above problem has a positive solution.
4. If $q = 2$ and $2\alpha \in [-1, 0)$ we prove some multiplicity results. Note that these results improve the multiplicity obtained in [4] for $2\alpha \geq 0$.

We will assume nonnegative solutions in all this part and then we will omit the explicit mention to this condition.

3.2.1 Some Preliminaries

We recall some known results that we use in this chapter. First the following comparison principle which is a consequence of the comparison results in [12] (see also [2]).

Lemma 3.1 *Let g be a nonnegative function such that $g \in L^1(\Omega)$, $s > 0$. Assume that w_1, w_2 are nonnegative functions such that $w_1, w_2 \in W_0^{1,p}(\Omega)$ $1 \le p < \dfrac{N}{N-1}$ verifying*

$$\begin{cases} -\Delta w_1 \le h(x)\dfrac{|\nabla w_1|^p}{1+s|\nabla w_1|^p} + g \text{ in } \Omega, \\ \qquad\qquad w_1 = 0 \text{ on } \partial\Omega. \end{cases} \quad \text{and} \quad \begin{cases} -\Delta w_2 \ge h(x)\dfrac{|\nabla w_2|^p}{1+s|\nabla w_2|^p} + g \text{ in } \Omega, \\ \qquad\qquad w_2 = 0 \text{ on } \partial\Omega. \end{cases} \tag{3.6}$$

where $h \in L^\infty(\Omega)$, then $w_2 \ge w_1$ in Ω.

We will also use the following result proved in the appendix of [20], that is the spirit of the arguments by Stampacchia in [138].

Lemma 3.2 *Assume that $u \in L^1_{loc}(\Omega)$ is such that $\Delta u \in L^1_{loc}(\Omega)$, then for all $p \in [0, \frac{N}{N-1})$, and for any open sets $\Omega_1 \subset \Omega_2 \subset \overline{\Omega}_2 \subset \Omega$, there exists a positive constant $C \equiv C(p, \Omega_1, \Omega_2, N)$ such that*

$$||u||_{W^{1,p}(\Omega_1)} \le C \int_{\Omega_2} (|u| + |\Delta u|)dx. \tag{3.7}$$

Moreover if $u \in L^1(\Omega)$ and $\Delta u \in L^1(\Omega)$, then the above estimate holds globally in the domain Ω.

Let us give an idea of the proof in the global case (valid for more general operator). The precise statement is as follows.

For any f belonging to $L^1(\Omega)$ there exists a unique $u \in L^1(\Omega)$ the distributional solution to

$$\begin{cases} -\Delta u = f \ \text{ in } \Omega \\ \quad u = 0 \ \text{ on } \partial\Omega, \end{cases} \tag{3.8}$$

that is,

$$\int_\Omega u(-\Delta\phi) = \int_\Omega f\phi$$

Moreover,

$$\forall k \geq 0 \qquad T_k(u) \in W^{1,2}(\Omega)\,, \tag{3.9}$$

$$u \in L^q\,, \qquad \forall\, q \in \Big(1, \frac{N}{N-2}\Big) \tag{3.10}$$

and

$$\big|\nabla u\big| \in L^r(\Omega)\,, \qquad \forall\, r \in \Big(1, \frac{N}{N-1}\Big)\,. \tag{3.11}$$

Uniqueness Assume that u is a weak solution to (3.8) with $f = 0$,

$$\int_\Omega u\psi \; dx = 0 \qquad \text{for any} \quad \psi \in \mathcal{C}_0^\infty(\Omega).$$

Therefore $u \equiv 0$.

Existence We obtain the solution to (3.8) as a limit of solutions to approximated problems.

Consider $f_n \in L^\infty(\Omega)$ such that $f_n \to f$ in $L^1(\Omega)$ and let u_n be the solution to the problem

$$\begin{cases} -\Delta u_n = f_n(x) & \text{in } \Omega,\\ u_n = 0 & \text{in } \partial\Omega. \end{cases} \tag{3.12}$$

Step 1 *There exists a positive constant c, only depending on N and Ω, such that*

$$\|u_n\|_{L^q} \leq c\, \|f_n\|_{L^1(\Omega)}, \qquad \forall q \in \Big(1, \frac{N}{N-2}\Big)\,. \tag{3.13}$$

Let us multiply the equation in (3.12) by $T_k(u_n)$, for $k \geq 0$, defined in (1.13), and let us integrate over Ω. We obtain by using the Sobolev inequality that

$$\|T_k(u_n)\|^2_{L^{2^*}} \leq \mathcal{S}^2\, k\|f_n\|_{L^1(\Omega)} \quad 2^* = \begin{cases} \frac{2N}{N-2} & \text{if } N > 2\\ \infty & \text{if } N = 1, 2 \end{cases}\,. \tag{3.14}$$

Therefore

$$k^2\, |A_{n,k}(u_n)|^{\frac{N-2}{N}} \leq \|T_k(u_n)\|^2_{L^{2^*_s}} \leq \mathcal{S}^2\, k\|f_n\|_{L^1(\Omega)}\,,$$

where $A_{n,k}(u_n) = \{x \in \Omega : u_n(x) \geq k\}$. It follows that

$$|A_{n,k}(u_n)| \leq c\Big(\frac{\|f_n\|_{L^1(\Omega)}}{k}\Big)^{\frac{N}{N-2}}. \tag{3.15}$$

It means that u_n is bounded in the Marcinkiewicz space $\mathcal{M}^{\frac{N}{N-2}}(\Omega)$ and consequently (3.13) holds true.

Step 2 *There exists a positive constant c, just depending on q, N, and* Ω, *such that*

$$\|\nabla u_n\|_{L^r(\Omega)} \leq c\,\|f_n\|_{L^1(\Omega)}, \qquad \forall r \in \Big(1, \frac{N}{N-1}\Big). \tag{3.16}$$

We fix $\lambda > 0$, and, for any positive k, we want to estimate the measure of the following set:

$$\begin{aligned}&\big\{x \in \Omega : |\nabla u_n| \geq \lambda\big\}\\ &= \big\{x \in \Omega : |\nabla u_n| \geq \lambda, u_n < k\big\} \cup \big\{x \in \Omega : |\nabla u_n| \geq \lambda, u_n \geq k\big\},\end{aligned}$$

and consequently

$$\big\{x \in \Omega : |\nabla u_n| \geq \lambda\big\} \subset \big\{x \in \Omega : |\nabla u_n| \geq \lambda, u_n < k\big\} \cup A_{n,k}(u_n).$$

Since

$$|\big\{x \in \Omega : |\nabla u_n| \geq \lambda,\ u_n < k\big\}| \leq \frac{1}{\lambda^2} \int_{\{x\in\Omega, u_n<k\}} |\nabla u_n|^2\,dx,$$

then

$$\begin{aligned}&|\big\{x \in \Omega : |\nabla u_n| \geq \lambda, u_n < k\}|\\ &\leq \frac{1}{\lambda^2} \int_\Omega |\nabla T_k(u_n)|^2\,dx \leq \frac{k}{\lambda^2}\|f_n\|_{L^1(\Omega)}.\end{aligned}$$

Moreover, using (3.15), we have that for every $k > 0$,

$$|\big\{x \in \Omega : |\nabla u_n| \geq \lambda\big\}| \leq \frac{k}{\lambda^2}\|f_n\|_{L^1(\Omega)} + c\Big(\frac{\|f_n\|_{L^1(\Omega)}}{k}\Big)^{\frac{N}{N-2}}.$$

Minimizing in k we find that the minimum is achieved by $k = \lambda^{\frac{N-2}{N-1}}\,\|f_n\|_{L^1(\Omega)}^{\frac{1}{N-1}}$, thus we have

$$|\big\{x \in \Omega : |\nabla u_n| \geq \lambda\big\}| \leq c\Big(\frac{\|f_n\|_{L^1(\Omega)}}{\lambda}\Big)^{\frac{N}{N-1}}. \tag{3.17}$$

This means that $|\nabla u_n|$ is bounded in the Marcinkiewicz space $\mathcal{M}^{\frac{N}{N-1}}(\Omega)$ and consequently (3.16) holds true.

Step 3 *Passing to the limit.* Before passing to the limit in the equation, we need to determine the a.e. limit of u_n. Using the linearity of the equation, we have that for any m and $n \in \mathbb{N}$, then $u_n - u_m$ solves

$$\begin{cases} -\Delta(u_n - u_m) = f_n - f_m & \text{in } \Omega, \\ u_n = 0, \quad u_m = 0 & \text{in } \partial\Omega. \end{cases}$$

Hence, choosing for any $k > 0$, $T_k(u_n - u_m)$ as a test function in the weak formulation of the above problem, we deduce, by repeating the computations of Step 1, that

$$|\{x \in \Omega : |u_n - u_m| \geq k\}| \leq c\Big(\frac{\|f_n - f_m\|_{L^1(\Omega)}}{k}\Big)^{\frac{N}{N-2}}.$$

Since the right hand side of the above inequality is small for n and m large enough, it follows that $\{u_n\}$ is a Cauchy sequence in measure. Consequently, up to subsequences, it converges in Ω almost every where, to a function u.

By the Step 1 we also deduce (using the embedding of the $\mathcal{M}^p(\Omega)$ spaces into the $L^p(\Omega)$, for $p \geq 1$) that u_n also converges to u in $L^q(\Omega)$, for any $1 \leq q < \frac{N}{N-2}$. Notice that this is sufficient to pass to the limit in the equation and obtain a weak solution of (3.8). Observe that, by the uniqueness, the whole sequence converges to u in $L^q(\Omega)$ and that (3.10) holds.

Since (3.17) holds, we have that

$$\big|\big\{x \in \Omega : |\nabla (u_n - u_m)| \geq \lambda\big\}\big| \leq c\Big(\frac{\|f_n - f_m\|_{L^1(\Omega)}}{\lambda}\Big)^{\frac{N}{N-s}},$$

thus ∇u_n is a Cauchy sequence in measure in Ω, therefore, up to a subsequence, ∇u_n converges a.e in Ω. Hence by Fatou lemma (3.9) follows by (3.14). Again by Fatou lemma and (3.17), we also obtain (3.11).

Also the approximating sequence $\{u_n\}$ verifies that for all $k > 0$

$$T_k(u_n) \to T_k(u) \text{ as } n \to \infty \text{ strongly in } W_0^{1,2}(\Omega). \tag{3.18}$$

The linearity of the problem allow us to assume $f \geq 0$ and then we have $u_n \uparrow u$. Then the strong convergence result is in particular obtained in the next Proposition.

Proposition 3.3 *Assume* $\{u_n\}$ *a sequence bounded in* $W^{1,p}_{loc}(\Omega)$ *for some* $1 < p \leq 2$ *and such that:*

(i) $u_n \rightharpoonup u$ *weakly in* $W^{1,p}_{loc}(\Omega)$.
(ii) $u_n \leq u$
(iii) $\|\nabla T_k u_n\|_2 \leq C$.

Then

$$\|\nabla T_k u_n - \nabla T_k u\|_2 \to 0 \text{ as } n \to \infty.$$

See Lemma 5.2 of [113] (see also [2] and [12]).

The following weak Harnack inequality (see [44]) will be the tool to obtain some local estimates.

Lemma 3.4 *Let* $h \in L^\infty(\Omega)$ *be a nonnegative function and assume that* v *solves*

$$\begin{cases} -\Delta v = h(x) \text{ in } \Omega, \\ \quad v = 0 \text{ on } \partial\Omega, \end{cases}$$

then

$$\frac{v(x)}{\delta(x)} \geq c(\Omega) \int_\Omega h(x)\delta(x)\, dx, \text{ for all } x \in \Omega,$$

where $\delta(x) = dist(x, \partial\Omega)$.

Since we consider nonnegative data, we will always deal with nonnegative solutions.

3.2.2 Existence Results for L^∞ Data

We start by finding a family of local radial supersolutions in the whole $\mathbb{R}^N$ to the equation

$$-\Delta w = w^{q\alpha}|\nabla w|^q. \tag{3.19}$$

Up to scaling argument it is sufficient to take $B_1(0)$, the unit ball centered in zero and guess a supersolution of the form $w(r) = Ar^{-\beta}$ for $0 < r \leq 1$ and $A > 0$. Then w is a super-solution to the above problem in $B_1(0)$ if $\beta < N - 2$ and

$$A\beta(N-2-\beta)r^{-\beta-2} \geq A^{q\alpha+q}\beta^q r^{-q\alpha\beta-q(\beta+1)} \tag{3.20}$$

Hence

$$\beta + 2 \geq q\alpha\beta + q(\beta+1), \text{ that is, } \beta(q(\alpha+1)-1) \leq 2 - q.$$

Therefore, we obtain:

1. If $q(\alpha+1)\le 1$, namely if $q\alpha \le 1-q$, the above inequality holds for all β satisfying $0<\beta<N-2$. We can write (3.20) as
$$A^{1-q\alpha-q}(N-2-\beta)\ge \beta^{q-1}r^{\beta+2-q\alpha\beta-q(\beta+1)}.$$
Since $\beta+2-q\alpha\beta-q(\beta+1)\ge 0$, we can choose $A>0$ such that the above inequality holds. For a such A we have a super-solution in $B_1(0)$ and then in all $\mathbb{R}^N$.
2. If $q(\alpha+1)>1$, then β must satisfy
$$\beta<\min\{\frac{2-q}{q(\alpha+1)-1}, N-2\}.$$
If we take the corresponding $A=A(\beta)$, we find again a super-solution.

If $q=2$, then $\beta=0$ and we will take as a super-solution $w(x)=A|x-x_0|^{-\gamma}$, $\gamma>0$, $x_0\in\mathbb{R}^N\setminus\Omega$ and A small enough.

The result for bounded data is the following.

Theorem 3.5 *Assume a function $f\in L^\infty(\Omega)$ such that $f\gneqq 0$ and consider the problem*
$$\begin{cases}-\Delta u=u^{q\alpha}|\nabla u|^q+\lambda f(x) & \text{in } \Omega\\ u=0 & \text{on } \partial\Omega,\end{cases}\tag{3.21}$$
$\Omega\subset\mathbb{R}^N$ a bounded domain.

Then,

1. *If $q(\alpha+1)\le 1$, problem (3.21) has a distributional bounded solution u such that $u^{q\alpha}|\nabla u|^q\in L^1_{loc}(\Omega)$ for any $\lambda>0$.*
2. *If $q(\alpha+1)>1$, the result holds true if λ is sufficiently small.*

Moreover

1. *$u^{\frac{\gamma}{q}+1}\in W_0^{1,q}(\Omega)$ with $\gamma>-\frac{q}{2-q}(\alpha q+1)$ if $q<2$, $\alpha q\le -1$*
2. *$u\in W_0^{1,q}(\Omega)$ if $q\le 2$, $-1<\alpha q<0$.*

In the case $q=2$, $2\alpha\le -1$ we have $\Psi(u)\in W_0^{1,2}(\Omega)$ where
$$\Psi(s)=\begin{cases}\displaystyle\int_0^s e^{\frac{t^{2\alpha+1}}{2\alpha+1}}dt & 2\alpha+1<0\\ \frac{s^2}{2} & 2\alpha+1=0\end{cases}\tag{3.22}$$

Remark 3.6 The regularity in particular shows how the boundary datum is attained, that if $\alpha q<0$ a priori is no clear. It is also worthy to point out that the techniques that we use in the proof, prove the existence result for $\alpha q=0$ $(m=1)$ studied in Sect. 1 for $q=2$.

Proof We divide the proof in two parts.

First Part $q(\alpha + 1) \leq 1$. It is important to notice that in this case there is no smallness condition on $||f||_\infty$. Let $B_R(0)$ be such that $\overline{\Omega} \subset B_R(0)$ and $0 < \beta < N - 2$. Let us consider the function

$$w(r) = Ar^{-\beta}, \ A > 0.$$

According with the previous computation, there exists $A = A(\beta)$ large enough for which the following inequality holds

$$(N - 2 - \beta) \geq A^{q(\alpha+1)-1}\beta^{q-1}r^{\beta+2-\beta\alpha q-q(\beta+1)} + \frac{||f||_\infty}{\beta A}.$$

which implies that $w(r)$ is an unbounded supersolution of our problem. We can easily construct a bounded super-solution $w_1(r)$ for f fixed, for example shifting the pole of the previous super-solution w to $z \in \mathbb{R}^N \backslash \overline{B_R(0)})$.

First Step We claim that for all $\varepsilon > 0$, the problem

$$\begin{cases} -\Delta u_\varepsilon = \dfrac{|\nabla u_\varepsilon|^q}{(u_\varepsilon + \varepsilon)^a} + f(x) & \text{in } \Omega \\ u_\varepsilon = 0 & \text{on } \partial\Omega, \end{cases} \tag{3.23}$$

has a bounded minimal solution u_ε such that $\dfrac{|\nabla u_\varepsilon|^q}{(u_\varepsilon + \varepsilon)^a} \in L^1(\Omega)$.

We will consider separately the arguments according with the value of q:

$$1). \quad 1 < q < 2 \quad \text{and } 2). \quad q = 2.$$

For simplicity of tapping we set $a = -q\alpha > 0$.

Case 1 $1 < q < 2$.

For each $n \in \mathbb{N}$, consider the following approximated problem

$$\begin{cases} -\Delta u_{\varepsilon,n} = \Big(\dfrac{1}{(u_{\varepsilon,n} + \varepsilon)^a}\Big)\dfrac{|\nabla u_{\varepsilon,n}|^q}{1 + \frac{1}{n}|\nabla u_{\varepsilon,n}|^q} + f(x) & \text{in } \Omega \\ u_{\varepsilon,n} = 0 & \text{on } \partial\Omega. \end{cases} \tag{3.24}$$

By the Leray-Lions Theorem (see [102]), for ε and n fixed, there exists a unique $u_{\varepsilon,n} \in W_0^{1,2}(\Omega) \cap L^\infty(\Omega)$ solution to (3.24). Moreover, using the result of Lemma 3.1 (see too [7]), we obtain that $u_{\varepsilon,n} \leq w_1$ for all $n \in \mathbb{N}$ and $\varepsilon > 0$.

In order to reach a global estimate in $L^1(\Omega)$,

$$\|u_n\|_{L^1(\Omega)} \leq C, \quad \|\Delta u_n\|_{L^1(\Omega)} \leq C,$$

we use $(1+u_{\varepsilon,n})^s - 1$ as a test function in (3.24) where $0 < s << 1$; using the fact that $u_{\varepsilon,n} \le w$ and Hölder inequality, we obtain

$$\int_\Omega \frac{|\nabla u_{\varepsilon,n}|^2}{(u_{\varepsilon,n}+1)^{1-s}}dx \le C(\varepsilon) \tag{3.25}$$

and

$$\int_\Omega \frac{(u_{\varepsilon,n}+1)^s}{(u_{\varepsilon,n}+\varepsilon)^a} \frac{|\nabla u_{\varepsilon,n}|^q}{1+\frac{1}{n}|\nabla u_{\varepsilon,n}|^q}dx \le C(\varepsilon). \tag{3.26}$$

Thus, again by Hölder inequality

$$||u_{\varepsilon,n}||_1 \le C, \qquad ||\Delta u_{\varepsilon,n}||_1 \le C(\varepsilon).$$

Hence, using Lemma 3.2 there exists a constant $C(\varepsilon)$, independent of n, such that

$$||u_{\varepsilon,n}||_{W_0^{1,p}(\Omega)} \le C(\varepsilon), \text{ for all } p < \frac{N}{N-1}, \text{ for all } n \in \mathbb{N}.$$

Therefore, there exists $u_\varepsilon \in W_0^{1,p}(\Omega)$, such that, up to a subsequence,

$$u_{\varepsilon,n} \rightharpoonup u_\varepsilon \text{ weakly in } W_0^{1,p}(\Omega) \text{ as } n \to +\infty.$$

We use the local compactness result in Proposition 3.3.
As consequence it results that, for all k and ε, as $n \to +\infty$

$$T_k(u_{\varepsilon,n}) \to T_k(u_\varepsilon) \text{ strongly in } W_0^{1,2}(\Omega).$$

From estimates (3.25) and (3.26), it follows that

$$\int_\Omega \frac{1}{(u_{\varepsilon,n}+\varepsilon)^a} \frac{|\nabla u_{\varepsilon,n}|^q}{1+\frac{1}{n}|\nabla u_{\varepsilon,n}|^q}dx$$
$$= \int_{u_{\varepsilon,n}<k} \frac{1}{(u_{\varepsilon,n}+\varepsilon)^a} \frac{|\nabla u_{\varepsilon,n}|^q}{1+\frac{1}{n}|\nabla u_{\varepsilon,n}|^q}dx + \int_{u_{\varepsilon,n}\ge k} \frac{1}{(u_{\varepsilon,n}+\varepsilon)^a} \frac{|\nabla u_{\varepsilon,n}|^q}{1+\frac{1}{n}|\nabla u_{\varepsilon,n}|^q}dx.$$

Notice that since (3.26) holds,

$$\int_{u_{\varepsilon,n}\ge k} \frac{1}{(u_{\varepsilon,n}+\varepsilon)^a} \frac{|\nabla u_{\varepsilon,n}|^q}{1+\frac{1}{n}|\nabla u_{\varepsilon,n}|^q}dx$$
$$\le \frac{1}{(1+k)^s} \int_{u_{\varepsilon,n}\ge k} \frac{(u_{\varepsilon,n}+1)^s}{(u_{\varepsilon,n}+\varepsilon)^a} \frac{|\nabla u_{\varepsilon,n}|^q}{(u_{\varepsilon,n}+1)^s}dx \le \frac{C(\varepsilon)}{(1+k)^s}.$$

Hence

$$\lim_{k\to\infty}\int_{u_{\varepsilon,n}\geq k}\frac{1}{(u_{\varepsilon,n}+\varepsilon)^a}\frac{|\nabla u_{\varepsilon,n}|^q}{1+\frac{1}{n}|\nabla u_{\varepsilon,n}|^q}dx=0,\ \text{uniformly in } n.$$

Thus, using Vitali's lemma we conclude that $u_\varepsilon \in W_0^{1,p}(\Omega)$ for any $p < \frac{N}{N-1}$ and $\frac{|\nabla u_\varepsilon|^q}{(u_\varepsilon+\varepsilon)^a} \in L^1(\Omega)$.

Hence u_ε is a bounded distributional solution to the problem

$$\begin{cases} -\Delta u_\varepsilon = \dfrac{|\nabla u_\varepsilon|^q}{(u_\varepsilon+\varepsilon)^a} + f(x) & \text{in } \Omega \\ u_\varepsilon = 0 & \text{on } \partial\Omega. \end{cases} \tag{3.27}$$

Thus the claim follows in this case.

Case 2 $q = 2$.

Now we have $a = -2\alpha > 1$. As above, by Leray-Lions Theorem in [102], we are able to solve the problem

$$\begin{cases} -\Delta u_{\varepsilon,n} = \left(\dfrac{1}{(u_{\varepsilon,n}+\varepsilon)^a}\right)\dfrac{|\nabla u_{\varepsilon,n}|^2}{1+\frac{1}{n}|\nabla u_{\varepsilon,n}|^2} + f(x) & \text{in } \Omega \\ u_{\varepsilon,n} = 0 & \text{on } \partial\Omega, \end{cases} \tag{3.28}$$

Let u_n be a solution. In order to get L^1 estimate on $u_{\varepsilon,n}$, it can be used the same argument as above. However we will perform a different approach to estimate $u_{\varepsilon,n}$ uniformly on n and ε. This argument will be useful later.

We start by proving a local estimate of $u_{\varepsilon,n}$. Take $\phi \in W_0^{1,2}(\Omega)$ the solution to $-\Delta\phi = 1$ and take $e^{-\frac{1}{(a-1)u_{\varepsilon,n}^{a-1}}}\phi$ as a test function in (3.28). It follows that

$$\int_\Omega \frac{\phi}{u_{\varepsilon,n}^a}e^{-\frac{1}{(a-1)u_{\varepsilon,n}^{a-1}}}|\nabla u_{\varepsilon,n}|^2dx + \int_\Omega K(u_{\epsilon,n})dx$$
$$\leq \int_\Omega e^{-\frac{1}{(a-1)u_{\varepsilon,n}^{a-1}}}\frac{\phi}{u_{\varepsilon,n}^a}|\nabla u_{\varepsilon,n}|^2dx + \int_\Omega e^{-\frac{1}{(a-1)u_{\varepsilon,n}^{a-1}}}\phi f dx$$

where $K(s) = \int_0^s e^{-\frac{1}{(a-1)t^{a-1}}}dt$.

Since $e^{-\frac{1}{(a-1)s^{a-1}}} \leq C$ and $K(s) \geq c_1 s - c_2$, hence we conclude that

$$\int_\Omega u_{\varepsilon,n}dx \leq C(1+\int_\Omega \phi f dx) \leq C(1+||f||_1).$$

Thus $||u_{\varepsilon,n}||_{L^1(\Omega)} \leq C$ where C is independent of n and ε.

Using now ϕ as a test function in (3.28) it results that

$$c\int_\Omega -\Delta u_{\varepsilon,n}\delta(x) \le c\int_\Omega -\Delta u_{\varepsilon,n}\phi(x) \le \int_\Omega \Big(\frac{1}{(u_{\varepsilon,n}+\varepsilon)^a}\frac{|\nabla u_{\varepsilon,n}|^q}{1+\frac{1}{n}|\nabla u_{\varepsilon,n}|^q}+f\Big)\phi\, dx$$

$$= \int_\Omega u_{\varepsilon,n}(-\Delta\phi) = \int_\Omega u_{\varepsilon,n} \le C \quad \text{for all } n.$$

Then $||\Delta u_{\varepsilon,n}||_{L^1(\Omega,\delta)} \le C$. Therefore, using Lemma 3.2, we get $||u_{\varepsilon,n}||_{W^{1,p}_{loc}(\Omega)} \le C$.

To reach a global estimate we define

$$J_\varepsilon(s) = \exp\{C(\frac{1}{\varepsilon^{a-1}} - \frac{1}{(s+\varepsilon)^{a-1}})\} - 1.$$

and use $J_\varepsilon(u_{\varepsilon,n})$ as a test function in (3.28). We obtain that

$$C(a-1)\int_\Omega \frac{(J_\varepsilon(u_{\varepsilon,n})+1)}{(u_{\varepsilon,n}+\varepsilon)^a}|\nabla u_{\varepsilon,n}|^2 dx + \int_\Omega \Big(\frac{1}{(u_{\varepsilon,n}+\varepsilon)^a}\Big)\frac{|\nabla u_{\varepsilon,n}|^2}{1+\frac{1}{n}|\nabla u_{\varepsilon,n}|^2}dx$$

$$\le \int_\Omega \frac{(J_\varepsilon(u_{\varepsilon,n})+1)}{(u_{\varepsilon,n}+\varepsilon)^a}|\nabla u_{\varepsilon,n}|^2 dx + \int_\Omega J_\varepsilon(u_{\varepsilon,n}) f dx.$$

Pick up C such that $C(a-1) > 1$, put $\gamma = C(a-1) - 1$ and take into account that $J_\varepsilon(s) \le C(\varepsilon)$. It follows that

$$\gamma\int_\Omega \frac{J_\varepsilon(u_{\varepsilon,n})+1}{(u_{\varepsilon,n}+\varepsilon)^a}|\nabla u_{\varepsilon,n}|^2 dx + \int_\Omega \frac{1}{(u_{\varepsilon,n}+\varepsilon)^a}\frac{|\nabla u_{\varepsilon,n}|^2}{1+\frac{1}{n}|\nabla u_{\varepsilon,n}|^2}dx \le C(\varepsilon)||f||_1.$$

Thus $||\Delta u_{\varepsilon,n}||_{L^1(\Omega)} \le C(\varepsilon)$. Therefore, as in the Case 1, there exists u_ε solution to problem

$$\begin{cases} -\Delta u_\varepsilon = \dfrac{|\nabla u_\varepsilon|^2}{(u_\varepsilon+\varepsilon)^a} + f(x) & \text{in } \Omega \\ \quad u_\varepsilon = 0 & \text{on } \partial\Omega. \end{cases} \tag{3.29}$$

Hence the claim follows.

Second Step Pass to the limit as $\varepsilon \to 0$.

By the construction of u_ε we have for free that $u_\varepsilon \le w_1$, the initial super-solution, and that the sequence $\{u_\varepsilon\}$ is increasing as $\varepsilon \downarrow 0$. Hence, by the result of Lemma 3.4 we obtain that

$$u_\varepsilon(x_0) \ge C\text{dist}(x_0, \partial\Omega)\int_\Omega \Big(\frac{|\nabla u_\varepsilon|^q}{(u_\varepsilon+\varepsilon)^a} + f(x)\Big)\text{dist}(x, \partial\Omega)dx,$$

for any $x_0 \in \Omega$. Since $u_\varepsilon \le w_1 \le C$ in Ω, it follows that

$$\int_\Omega \Big(\frac{|\nabla u_\varepsilon|^q}{(u_\varepsilon+\varepsilon)^a} + f(x)\Big)\mathrm{dist}(x,\partial\Omega)dx \le C \text{ for all } \varepsilon.$$

Thus, since $u_\varepsilon \le w_1$, where w_1 is the super-solution defined at the beginning of the proof, we have $||u_\varepsilon||_{L^1(\Omega)} \le C$ and $||\Delta u_\varepsilon||_{L^1_{loc}(\Omega)} \le C$. Hence from Lemma 3.2 there exists a constant C independent of ε such that

$$||u_\varepsilon||_{W^{1,p}_{loc}(\Omega)} \le C \text{ for all } 1 < p < \frac{N}{N-1}.$$

Since $\{u_\varepsilon\}$ is uniformly bounded in $L^\infty(\Omega)$, then, using a variation of the local compactness result obtained in Proposition 3.3, it results that for all $k > 0$

$$T_k(u_\varepsilon) \to T_k(u) \text{ strongly in } W^{1,2}_{loc}(\Omega) \text{ as } \epsilon \to 0.$$

Let $\phi \in C_0^\infty(\Omega)$ and consider $(1+u_\varepsilon)^s\phi$ as a test function in (3.23). As above, by Hölder inequality it results that, for all ε

$$\int_\Omega \frac{|\nabla u_\varepsilon|^2}{(u_\varepsilon+1)^{1-s}}\phi dx \le C \tag{3.30}$$

and

$$\int_\Omega \phi\frac{(u_\varepsilon+1)^s}{(u_\varepsilon+\varepsilon)^a}|\nabla u_\varepsilon|^q dx \le C. \tag{3.31}$$

Hence we get the compactness in $L^1_{loc}(\Omega)$ of the sequence $\dfrac{|\nabla u_\varepsilon|^q}{(u_\varepsilon+\varepsilon)^a}$ and then the existence result follows.

Second Part $q(\alpha+1) > 1$.

Consider the following approximated problem

$$\begin{cases} -\Delta u_{\varepsilon,n} = (u_{\varepsilon,n}+\varepsilon)^{\alpha q}\dfrac{|\nabla u_{\varepsilon,n}|^q}{1+\frac{1}{n}|\nabla u_{\varepsilon,n}|^q} + \lambda f(x) & \text{in } \Omega \\ u_{\varepsilon,n} = 0 & \text{on } \partial\Omega, \end{cases} \tag{3.32}$$

if $\alpha q < 0$, or

$$\begin{cases} -\Delta u_{\varepsilon,n} = T_{\frac{1}{\varepsilon}}(u_{\varepsilon,n}^{\alpha q})\dfrac{|\nabla u_{\varepsilon,n}|^q}{1+\frac{1}{n}|\nabla u_{\varepsilon,n}|^q} + \lambda f(x) & \text{in } \Omega \\ u_{\varepsilon,n} = 0 & \text{on } \partial\Omega, \end{cases} \tag{3.33}$$

if $\alpha q \ge 0$.

Since $q(\alpha+1) > 1$ and $f \in L^\infty(\Omega)$, in order to find a super-solution, we need to have that $\lambda \le \lambda_0$ where λ_0 is a fixed constant that depends on β and $||f||_\infty$.

Notice that in this case we can choose as a super-solution the function $w(x) = A|x-z|^{-\gamma}$, with $z \in \mathbb{R}^N\backslash\overline{\Omega}$, γ any positive constant and A small enough.

Therefore following similar arguments as in the First Part, we prove the existence of a bounded distributional solution u to problem (3.33) with $u^{\alpha q}|\nabla u|^q \in L^1_{loc}(\Omega)$.

Regularity

(I) If $q\alpha > 0$ we have $u \in W_0^{1,2}(\Omega)$ as similar arguments to the one in Sect. 1.

(II) If $q\alpha < 0$ we distinguish the following cases.

i) $1 < q < 2$

a) $\alpha q \le -1$
b) $-1 < \alpha q < 0$

ii) $q = 2$

To reach the regularity result for $q < 2$ we will consider two cases:

(II) i) a) Assume that $\alpha q \le -1$. In this we have $a \ge 1$. Let u_ε the minimal solution to problem (3.23), consider $0 < \theta < \gamma$, to be chosen later. Then by Young inequality, for $\rho > 0$,

$$\int_\Omega u_\varepsilon^\gamma |\nabla u_\varepsilon|^q dx = \int_\Omega u_\varepsilon^\theta |\nabla u_\varepsilon|^q u_\varepsilon^{\gamma-\theta} dx$$
$$\le \rho \int_\Omega u_\varepsilon^{\frac{2\theta}{q}} |\nabla u_\varepsilon|^2 dx + C(\rho) \int_\Omega u_\varepsilon^{\frac{2(\gamma-\theta)}{2-q}} dx$$

Now using $u_\varepsilon^{\frac{2\theta}{q}+1}$ as a test function in (3.27) and using the fact that $u_\varepsilon \le w_1 \le C$ in Ω, it follows that

$$(\frac{2\theta}{q}+1)\int_\Omega u_\varepsilon^{\frac{2\theta}{q}} |\nabla u_\varepsilon|^2 dx \le \int_\Omega u_\varepsilon^{\frac{2\theta}{q}+1-a} |\nabla u_\varepsilon|^q dx + \int_\Omega w_1^{\frac{2\theta}{q}+1-a} f.$$

Pick up $\gamma = \frac{2\theta}{q} + 1 - a$, that is $\theta = \frac{q}{2}(\gamma + a - 1)$. We need $(\gamma - \theta)\frac{2}{2-q} > 0$, therefore it is sufficient that

$$\gamma > -\frac{2}{2-q}(a-1).$$

Choosing ρ small enough and using again the fact that $u_\varepsilon \le w_1 \le C$, we have

$$\int_\Omega u_\varepsilon^\gamma |\nabla u_\varepsilon|^q dx \le C + C(\theta)\int_\Omega u_\varepsilon^{\frac{2(\gamma-\theta)}{2-q}} dx \le C.$$

Hence $u_\varepsilon^{1+\frac{\gamma}{q}}$ is bounded in $W_0^{1,q}(\Omega)$ and then $u_\varepsilon^{1+\frac{\gamma}{q}} \rightharpoonup u^{1+\frac{\gamma}{q}}$ weakly in $W_0^{1,q}(\Omega)$.

(II) i) b) Assume $-1 < \alpha q < 0$. In this case we have $a \in (0,1)$.

Using the computation above with $\gamma = 0$, it follows that

$$\int_\Omega |\nabla u_\varepsilon|^q dx \le \rho \int_\Omega u_\varepsilon^{\frac{2\theta}{q}} |\nabla u_\varepsilon|^2 dx + C(\theta)\int_\Omega u_\varepsilon^{\frac{-2\theta}{2-q}} dx.$$

Choosing $u_\varepsilon^{\frac{2\theta}{q}+1}$ as a test function in (3.27), with $\theta > 0$ such that $\frac{-2\theta}{2-q} \ge -1$, we get

$$(\frac{2\theta}{q}+1)\int_\Omega u_\varepsilon^{\frac{2\theta}{q}} |\nabla u_\varepsilon|^2 dx \le \int_\Omega u_\varepsilon^{\frac{2\theta}{q}+1-a} |\nabla u_\varepsilon|^q dx + \int_\Omega f u_\varepsilon^{\frac{2\theta}{q}+1} dx.$$

Notice that since $a < 1$, and $\theta > 0$, the condition $\frac{2\theta}{q} + 1 - a \ge 0$ holds.

Therefore using the fact that $C\mathrm{dist}(x, \partial\Omega) \le u_\varepsilon \le C_1$ and choosing ρ small, we conclude that

$$C\int_\Omega |\nabla u_\varepsilon|^q dx \le C + C\int_\Omega (\mathrm{dist}(x,\partial\Omega))^{\frac{-2\theta}{2-q}} dx \le C.$$

Hence $u_\epsilon \rightharpoonup u$ weakly in $W_0^{1,q}(\Omega)$.

(II) ii) $q = 2$. Recall the definition of function $\Psi(s)$ given in (3.22)

$$\Psi(s) = \begin{cases} \displaystyle\int_0^s e^{\frac{t^{2\alpha+1}}{2\alpha+1}} dt & 2\alpha+1 < 0 \\ \frac{s^2}{2} & 2\alpha+1 = 0. \end{cases}$$

Take $v_\epsilon \equiv e^{\frac{u_\varepsilon^{1-a}}{1-a}} \Psi(u_\varepsilon)$ as a test function in (3.23). It follows that

$$\int_\Omega |\nabla \Psi(u_\varepsilon)|^2 dx = \int_\Omega e^{\frac{2u_\varepsilon^{1-a}}{1-a}} |\nabla u_\varepsilon|^2 dx$$

$$\le \int_\Omega \langle \nabla u_\varepsilon, \nabla v_\varepsilon\rangle = \int_\Omega \frac{v_\varepsilon |\nabla u_2|^2}{(u_\varepsilon+\varepsilon)^a} + \int_\Omega f v_e$$

$$\leq \int_\Omega f e^{\frac{u_\varepsilon^{1-a}}{1-a}} \Psi(u_\varepsilon)dx$$

$$\leq C \int_\Omega \Psi(u_\varepsilon)dx.$$

Hence $||\Psi(u_\varepsilon)||_{W_0^{1,2}(\Omega)} \leq C$ and then $\Psi(u_\varepsilon) \rightharpoonup \Psi(u)$ weakly in $W_0^{1,2}(\Omega)$.

□

It is clear that the same existence result holds if we assume that $|x|^{(\beta+2)} f \in L^\infty(\Omega)$ by choosing a suitable β to have radial super-solution.

The exponent $q(\alpha+1)$ appearing in Theorem 3.5 is natural to construct a radial local super-solution w in the whole space. This exponent becomes a threshold for the behavior of the solutions to the problem.

3.2.3 Existence for $q\alpha < -1$ and General Datum in $L^1(\Omega)$

A remarkable fact is that for $q\alpha < -1$ we are able to find a distributional solution to problem (3.21) with data in $L^1(\Omega)$ and without any restriction in the size. In this setting we must to change the method of monotonicity to find a priori estimates, including the compactness arguments to be used. The main result of this section is the following.

Theorem 3.7 *Assume $f \in L^1(\Omega)$ such that $f \gneqq 0$. Suppose that $1 < q \leq 2$ and $q\alpha < -1$, then problem (3.21) has a distributional solution u such that $u^{q\alpha}|\nabla u|^q \in L^1_{loc}(\Omega)$, $u \in W^{1,p}_{loc}(\Omega)$ with $p < \frac{N}{N-1}$, and $\bar{\Psi}(u) \in W_0^{1,2}(\Omega)$ where*

$$\bar{\Psi}(s) = \int_0^s t^{\frac{q\alpha}{2}} e^{-\frac{\sigma}{2}t^{q\alpha+1}} dt, \text{ with } \sigma > \frac{1}{-q\alpha-1}.$$

Proof As above we set $-q\alpha = a$, then $a > 1$. We proceed by solving the approximated problems

$$\begin{cases} -\Delta u_n = \left(\dfrac{1}{(u_n+\frac{1}{n})^a}\right)\dfrac{|\nabla u_n|^q}{1+\frac{1}{n}|\nabla u_n|^q} + T_n(f) \text{ in } \Omega \\ u_n = 0 \qquad\qquad\qquad\qquad \text{on } \partial\Omega. \end{cases} \tag{3.34}$$

The existence of a solution u_n is a consequence of the Leray-Lions Theorem. Consider $\phi \in W_0^{1,2}(\Omega)$ the solution to

$$-\Delta\phi = 1 \quad \text{in } \Omega$$

and call

$$K(s)=\int_0^s \mathrm{e}^{\left(\frac{-1}{(a-1)t^{a-1}}\right)}dt.$$

Using $\mathrm{e}^{-\frac{1}{(a-1)u_n^{a-1}}}\phi$ as a test function in (3.34), it follows that

$$\begin{aligned}
\int_\Omega \langle\nabla u_n,\nabla v_n\rangle &= \int_\Omega \frac{\phi}{u_n^a}\mathrm{e}^{-\frac{1}{(a-1)u_n^{a-1}}}|\nabla u_n|^2+\int_\Omega \langle \mathrm{e}^{-\frac{1}{(a-1)u_n^{a-1}}}\nabla u_n,\nabla\phi\rangle\\
&=\int_\Omega \langle\nabla K(u_n),\nabla\phi\rangle+\int_\Omega \frac{\phi}{u_n^a}\mathrm{e}^{-\frac{1}{(a-1)u_n^{a-1}}}|\nabla u_n|^2\\
&=\int_\Omega K(u_n)+\int_\Omega \frac{\phi}{u_n^a}\mathrm{e}^{-\frac{1}{(a-1)u_n^{a-1}}}|\nabla u_n|^2\\
&=\int_\Omega \mathrm{e}^{-\frac{1}{(a-1)u_n^{a-1}}}\phi\Big(\frac{1}{(u_n+\frac{1}{n})^a}\Big)\frac{|\nabla u_n|^q}{1+\frac{1}{n}|\nabla u_n|^q}+\int_\Omega T_n(f)\mathrm{e}^{-\frac{1}{(a-1)u_n^{a-1}}}\phi\\
&\le \int_\Omega\int_\Omega e^{-\frac{1}{(a-1)u_n^{a-1}}}\frac{\phi}{u_n^a}|\nabla u_n|^q dx+\int_\Omega e^{-\frac{1}{(a-1)u_n^{a-1}}}\phi T_n(f)dx,
\end{aligned}$$

that we can summarize as

$$\begin{aligned}
&\int_\Omega e^{-\frac{1}{(a-1)u_n^{a-1}}}|\nabla u_n|^2\frac{\phi}{u_n^a}dx+\int_\Omega K(u_n)dx\\
&\quad\le \int_\Omega e^{-\frac{1}{(a-1)u_n^{a-1}}}\frac{\phi}{u_n^a}|\nabla u_n|^q dx+\int_\Omega e^{-\frac{1}{(a-1)u_n^{a-1}}}\phi T_n(f)dx
\end{aligned}$$

Since $\frac{1}{s^a}e^{-\frac{1}{(a-1)s^{a-1}}}\le C$ and $K(s)\ge c_1 s-c_2$, hence we conclude that if $q=2$,

$$\int_\Omega u_n dx\le c+\int_\Omega \phi f dx\le C(1+||f||_1).$$

If $q<2$, using Young inequality,

$$\begin{aligned}
&\int_\Omega \frac{1}{u_n^a}e^{-\frac{1}{(a-1)u_n^{a-1}}}|\nabla u_n|^2\phi dx+\int_\Omega K(u_n)dx\\
&\quad\le \int_\Omega e^{-\frac{1}{(a-1)u_n^{a-1}}}\frac{1}{u_n^a}|\nabla u_n|^2\phi dx+C\int_\Omega e^{-\frac{1}{(a-1)u_n^{a-1}}}\frac{1}{u_n^a}\phi dx+\int_\Omega e^{-\frac{1}{(a-1)u_n^{a-1}}}\phi f dx.
\end{aligned}$$

Thus in both cases we obtain that

$$\int_\Omega u_n dx\le C.$$

Now using ϕ as a test function in (3.34) it results that

$$C \geq \int_\Omega u_n = \int_\Omega \Big(\Big(\frac{1}{(u_n+\frac{1}{n})^a}\Big)\frac{|\nabla u_{\varepsilon,n}|^q}{1+\frac{1}{n}|\nabla u_{\varepsilon,n}|^q} + T_n(f)\Big)\phi dx \leq C \text{ for all } n. \tag{3.35}$$

As a consequence $||\Delta u_n||_{L^1(\Omega,\delta)} \leq C$ and thus from the result of Lemma 3.2 we get $||u_n||_{W^{1,p}_{loc}} \leq C$. Using $T_k(u_n)\phi$ as a test function in (3.34), we obtain that

$$\int_\Omega |\nabla T_k(u_n)|^2 \phi dx \leq Ck.$$

Therefore there exists $u \in W^{1,p}_{loc}(\Omega)$, for any $p < \frac{N}{N-1}$, such that $u_n \rightharpoonup u$ weakly in $W^{1,p}_{loc}(\Omega)$ and

$$T_k(u_n) \rightharpoonup T_k(u) \text{ weakly in } W^{1,2}_{loc}(\Omega).$$

To prove the summability that justify how the data is attained we proceed as follows. For $\sigma > \frac{1}{a-1}$, consider $e^{-\frac{\sigma}{u_n^{a-1}}}$ as a test function in (3.34), it follows that

$$(\sigma(a-1)-1)\int_\Omega \frac{e^{-\frac{\sigma}{u_n^{a-1}}}}{u_n^a}|\nabla u_n|^2 dx \leq \int_\Omega T_n(f)e^{\frac{u_n^{1-a}}{1-a}}dx \leq \int_\Omega T_n(f)dx \leq C \quad \text{if } q=2,$$

and

$$\sigma(a-1)\int_\Omega \frac{e^{-\frac{\sigma}{u_n^{a-1}}}}{u_n^a}|\nabla u_n|^2 dx \leq \int_\Omega \frac{e^{-\frac{\sigma}{u_n^{a-1}}}}{u_n^a}|\nabla u_n|^q dx + \int_\Omega T_n(f)e^{-\frac{\sigma}{u_n^{a-1}}}dx \text{ if } q<2.$$

In the case where $q < 2$, using again the Young inequality and for θ small enough, we obtain that

$$(\sigma(a-1)-\theta)\int_\Omega \frac{e^{-\frac{\sigma}{u_n^{a-1}}}}{u_n^a}|\nabla u_n|^2 dx \leq C(\theta)\int_\Omega \frac{e^{-\frac{\sigma}{u_n^{a-1}}}}{u_n^a}dx + \int_\Omega T_n(f)dx.$$

Hence in both cases we conclude that $||\bar{\Psi}(u_n)||_{W_0^{1,2}(\Omega)} \leq C$ and then, up to a subsequence,

$$\bar{\Psi}(u_n) \rightharpoonup \bar{\Psi}(u) \text{ weakly in } W_0^{1,2}(\Omega).$$

Let now $0 \leq \psi \in C_0^\infty(\Omega)$ be fixed. By the comparison result of Lemma 3.1, there exits a positive constant C_ψ, independent of n, such that $u_n^s(x) \geq C_\psi$ for all $x \in$

$\text{supp}(\psi)$; hence we can use $\dfrac{\psi}{u_n^s}$ as a test function in (3.34) where s is positive and sufficiently small. In this way we obtain

$$\begin{aligned}&-s\int_\Omega \frac{|\nabla u_n|^2}{u_n^{s+1}}\psi dx + \int_\Omega \frac{\nabla\psi}{u_n^s}\nabla u_n dx\\&= \int_\Omega \Big(\frac{1}{(u_n+\frac{1}{n})^a}\Big)\frac{|\nabla u_n|^q}{1+\frac{1}{n}|\nabla u_n|^q}\frac{\psi}{u_n^s}dx + \int_\Omega T_n(f)\frac{\psi}{u_n^s}dx,\end{aligned}$$

that is,

$$\begin{aligned}&\frac{1}{1-s}\int_\Omega u_n^{1-s}(-\Delta\psi)dx\\&= s\int_\Omega \frac{|\nabla u_n|^2}{u_n^{s+1}}\psi dx + \int_\Omega \Big(\frac{1}{(u_n+\frac{1}{n})^a}\Big)\frac{|\nabla u_n|^q}{1+\frac{1}{n}|\nabla u_n|^q}\frac{\psi}{u_n^s}dx + \int_\Omega T_n(f)\frac{\psi}{u_n^s}dx.\end{aligned}$$

Since $\{u_n\}$ is bounded in $L^1(\Omega)$ and $u_n^s(x) \geq C_\psi$ for all $x \in \text{supp}(\psi)$, then

$$\int_\Omega \frac{|\nabla u_n|^2}{u_n^{s+1}}\psi dx \leq C \text{ for all } n. \tag{3.36}$$

To get the local strong convergence of $T_k(u_n)$ we follow closely the argument used in [101]. We begin by considering the case $q < 2$.

Fixed $0 \leq \psi \in \mathcal{C}_0^\infty(\Omega)$ and let $h > k > 0$ to be chosen later. Define

$$w_n = T_{2k}(u_n - T_h(u_n) + T_k(u_n) - T_k(u)).$$

It is clear that $w_n \in W^{1,2}_{loc}(\Omega)$ and $\nabla w_n \equiv 0$ for $u_n > M \equiv 4k + h$. Using $w_n\psi$ as a test function in (3.34) it follows that

$$\int_\Omega \nabla T_M(u_n)\nabla w_n\psi dx + \int_\Omega w_n\nabla u_n\nabla\psi dx \leq \int_\Omega |w_n|\psi\frac{|\nabla u_n|^q}{u_n^a}dx + \int_\Omega |w_n|T_n(f)\psi dx.$$

Notice that

$$\begin{aligned}\int_\Omega \nabla T_M(u_n)\nabla w_n\psi dx &= \int_{u_n\leq k}\nabla T_k(u_n)\nabla w_n\psi dx + \int_{u_n>k}\nabla T_M(u_n)\nabla w_n\psi dx\\&\geq \int_\Omega \nabla T_k(u_n)\nabla(T_k(u_n)-T_k(u))\psi dx - \int_{u_n>k}|\nabla T_M(u_n)||\nabla T_k(u)|\psi dx,\end{aligned}$$

where the last integral is estimated as follows

$$
\begin{aligned}
\int\limits_{u_n>k} \nabla T_M(u_n)\nabla w_n\psi\,dx &= \int\limits_{h>u_n>k} \nabla T_M(u_n)\nabla w_n\psi\,dx + \int\limits_{u_n>h} \nabla T_M(u_n)\nabla w_n\psi\,dx\\
&= \int\limits_{h>u_n>k} \nabla T_M(u_n)\nabla T_{2k}(k-T_k(u))\psi\,dx + \int\limits_{h<u_n<M} \nabla T_M(u_n)\nabla T_{2k}(u_n-h+k-T_k(u))\psi\,dx\\
&= -\int\limits_{h>u_n>k} \nabla T_M(u_n)\nabla T_k(u)\psi\,dx + \int\limits_{h<u_n<h+k+T_k(u)} \nabla T_M(u_n)\nabla(u_n-h+k-T_k(u))\psi\,dx\\
&\geq -\int\limits_{u_n>k} |\nabla T_M(u_n)||\nabla T_k(u)|\psi\,dx.
\end{aligned}
\tag{3.37}
$$

Thus

$$
\begin{aligned}
&\int_\Omega |\nabla T_k(u_n)-\nabla T_k(u)|^2\psi\,dx\\
&\quad= \int_\Omega \nabla T_k(u_n)(\nabla T_k(u_n)-\nabla T_k(u))\psi\,dx - \int_\Omega \nabla T_k(u)(\nabla T_k(u_n)-\nabla T_k(u))\psi\,dx\\
&\quad\leq \int_\Omega \nabla T_M(u_n)\nabla w_n\psi\,dx + \int_{u_n>k} |\nabla T_M(u_n)||\nabla T_k(u)|\psi\,dx\\
&\quad\quad- \int_\Omega \nabla T_k(u)(\nabla T_k(u_n)-\nabla T_k(u))\psi\,dx.
\end{aligned}
$$

We analyze each term in the previous inequality.

Since $T_k(u_n) \rightharpoonup T_k(u)$ as $n\to\infty$, weakly in $W^{1,2}_{loc}(\Omega)$, it follows that

$$
\int_\Omega \nabla T_k(u)(\nabla T_k(u_n)-\nabla T_k(u))\psi\,dx = o(1)
$$

On the other hand, since $\chi_{\{u_n>k\}}|\nabla T_k(u)| \to 0$ strongly in $L^2_{loc}(\Omega)$ as $n\to\infty$, by using the fact that $|\nabla T_M(u_n)|$ is bounded in L^2_{loc}, then it results that

$$
\int_{u_n>k} |\nabla T_M(u_n)||\nabla T_k(u)|\psi\,dx \to 0 \text{ as } n\to\infty.
$$

Therefore we obtain that

$$
\begin{aligned}
\int_\Omega |\nabla T_k(u_n)-\nabla T_k(u)|^2\psi\,dx &\leq \int_\Omega \nabla T_M(u_n)\nabla w_n\psi\,dx + o(1)\\
&\leq \int_\Omega |w_n|\psi\frac{|\nabla u_n|^q}{u_n^a}dx + \int_\Omega |w_n|T_n(f)\psi\,dx - \int_\Omega w_n\nabla u_n\nabla\psi\,dx
\end{aligned}
$$

Notice that $w_n \to T_{2k}(u - T_h(u))$ strongly in $L^\sigma_{loc}(\Omega)$ for all $\sigma > 1$. Since $\{u_n\}$ is bounded in $W^{1,p}_{loc}$ for all $p < \frac{N}{N-1}$, there result

$$\lim_{n\to\infty} \int_\Omega w_n \nabla u_n \nabla \psi dx = \int_\Omega T_{2k}(u - T_h(u)) \nabla u \nabla \psi dx$$

and

$$\lim_{n\to\infty} \int_\Omega |w_n| T_n(f) \psi = \int_\Omega |T_{2k}(u - T_h(u))| f \psi dx.$$

It is clear that $T_{2k}(u - T_h(u)) \to 0$ as $h \to \infty$ in L^∞ weak $*$ topology, for each k fixed; since $|\nabla u||\nabla \psi| \in L^1$, then for all ε there exists $h_1(\varepsilon) >> k$ such that for $h > h_1(\varepsilon)$, we have

$$\int_\Omega T_{2k}(u - T_h(u))|\nabla u||\nabla \psi| dx + \int_\Omega |T_{2k}(u - T_h(u))| f \psi dx \le \varepsilon.$$

We deal now with the term $\int_\Omega |w_n| \psi \frac{|\nabla u_n|^q}{u_n^a} dx$. Since $q < 2$, then using Hölder inequality and the uniform estimate obtained in (3.36), we get

$$\begin{aligned} \int_\Omega |w_n| \frac{|\nabla u_n|^q}{u_n^a} \psi dx &\le \Big(\int_\Omega \psi \frac{|\nabla u_n|^2}{u_n^a} \psi dx\Big)^{\frac{q}{2}} \Big(\int_\Omega \frac{|w_n|^{\frac{2}{2-q}}}{u_n^a} \psi dx\Big)^{\frac{2-q}{2}} \\ &\le C\Big(\int_\Omega |w_n|^{\frac{2}{2-q}} \psi dx\Big)^{\frac{2-q}{2}}. \end{aligned}$$

It is clear that

$$\int_\Omega |w_n|^{\frac{2}{2-q}} \psi dx \to \int_\Omega T_{2k}(u - T_h(u))^{\frac{2}{2-q}} \psi dx, \quad n \to \infty,$$

then we can chose $h_2(\varepsilon) >> k$ to get

$$\int_\Omega T_{2k}(u - T_h(u))^{\frac{2}{2-q}} \psi dx \le \varepsilon \text{ if } h \ge h_2(\varepsilon).$$

Thus for $h \ge \max\{h_1(\varepsilon), h_2(\varepsilon)\}$, there result that

$$\limsup_{n\to\infty} \int_\Omega |\nabla T_k(u_n) - \nabla T_k(u)|^2 \psi dx \le \varepsilon.$$

Hence $T_k(u_n) \to T_k(u)$ strongly in $W^{1,2}_{loc}(\Omega)$ and the result follows in this case.

If $q = 2$, define $v_n = e^{-\frac{1}{(a-1)u_n^{a-1}}} w_n \psi$, with

$$w_n = T_{2k}((u_n - T_h(u_n) + T_k(u_n) - T_k(u)))_+.$$

Using v_n as a test function in (3.34) we obtain that

$$\begin{aligned}&\int_\Omega e^{-\frac{1}{(a-1)u_n^{a-1}}}\nabla T_M(u_n)\nabla w_n\psi\,dx+\int_\Omega e^{-\frac{1}{(a-1)u_n^{a-1}}}w_n\nabla u_n\nabla\psi\,dx\\&+\int_\Omega\frac{1}{u_n^a}e^{-\frac{1}{(a-1)u_n^{a-1}}}|\nabla u_n|^2w_n\psi\,dx\\&\le\int_\Omega w_n\psi e^{-\frac{1}{(a-1)u_n^{a-1}}}\frac{|\nabla u_n|^2}{u_n^a}dx+\int_\Omega w_nT_n(f)\psi\,dx.\end{aligned}$$

Since $e^{-\frac{1}{(a-1)u_n^{a-1}}}\le 1$, cancelling similar terms and using the same estimate as in (3.37), it follows that, for $h\ge h(\varepsilon)>>k$,

$$\limsup_{n\to\infty}\left(\left|\int_\Omega e^{-\frac{1}{(a-1)u_n^{a-1}}}w_n\nabla u_n\nabla\psi\,dx\right|+\left|\int_\Omega w_nT_n(f)\psi\,dx\right|\right)\le\varepsilon.$$

Thus

$$\limsup_{n\to\infty}\left|\int_\Omega e^{-\frac{1}{(a-1)u_n^{a-1}}}\nabla T_M(u_n)\nabla w_n\psi\,dx\right|\le\varepsilon. \tag{3.38}$$

Let $B_n=\{x\in\Omega: u_n-T_h(u_n)+T_k(u_n)-T_k(u)\ge 0\}$, as above we have

$$\begin{aligned}&\int_{B_n}e^{-\frac{1}{(a-1)u_n^{a-1}}}\nabla T_M(u_n)\nabla w_n\psi\,dx\\&=\int_{\{u_n\le k\}\cap B_n}e^{-\frac{1}{(a-1)u_n^{a-1}}}\nabla T_k(u_n)\nabla w_n\psi\,dx\\&+\int_{\{u_n>k\}\cap B_n}e^{-\frac{1}{(a-1)u_n^{a-1}}}\nabla T_M(u_n)\nabla w_n\psi\,dx\\&\ge\int_{B_n}e^{-\frac{1}{(a-1)u_n^{a-1}}}\nabla T_k(u_n)\nabla(T_k(u_n)-T_k(u))\psi\,dx\\&-\int_{u_n>k}e^{-\frac{1}{(a-1)u_n^{a-1}}}|\nabla T_M(u_n)||\nabla T_k(u)|\psi\,dx.\end{aligned}$$

By using the estimate (3.37), it is clear that

$$\int_{u_n>k}e^{-\frac{1}{(a-1)u_n^{a-1}}}|\nabla T_M(u_n)||\nabla T_k(u)|\psi\,dx\to 0\text{ as }n\to\infty.$$

Thus, using (3.38), it follows that

$$\limsup_{n\to\infty}\int_{B_n}e^{-\frac{1}{(a-1)u_n^{a-1}}}\nabla T_k(u_n)\nabla(T_k(u_n)-T_k(u))\psi\,dx\le 0. \tag{3.39}$$

To deal with the set $\Omega\backslash B_n$, we take $w_n = T_{2k}((u_n - T_h(u_n) + T_k(u_n) - T_k(u)))_-$.

The same computation as above, choosing $h \geq h_1(\varepsilon) >> k$, allow us to conclude that

$$\limsup_{n\to\infty} \int_{\Omega\backslash B_n} e^{-\frac{1}{(a-1)u_n^{a-1}}} \nabla T_k(u_n)\nabla(T_k(u_n) - T_k(u))\psi dx \leq 0. \tag{3.40}$$

Therefore combining (3.39) and (3.40), we reach

$$\limsup_{n\to\infty} \int_{\Omega} e^{-\frac{1}{(a-1)u_n^{a-1}}} \nabla T_k(u_n)\nabla(T_k(u_n) - T_k(u))\psi dx \leq 0.$$

Notice that

$$\begin{aligned}
&\int_{\Omega} e^{-\frac{1}{(a-1)u_n^{a-1}}} \nabla T_k(u_n)\nabla(T_k(u_n) - T_k(u))\psi dx \\
&= \int_{\Omega} e^{-\frac{1}{(a-1)u_n^{a-1}}} |\nabla T_k(u_n) - \nabla T_k(u)|^2\psi dx \\
&- \textstyle\int_{\Omega} e^{-\frac{1}{(a-1)u_n^{a-1}}} \nabla T_k(u)\nabla(T_k(u_n) - T_k(u))\psi dx \\
&= \int_{\Omega} e^{-\frac{1}{(a-1)u_n^{a-1}}} |\nabla T_k(u_n) - \nabla T_k(u)|^2\psi dx + o(1)
\end{aligned}$$

Since $u_n(x) \geq C_\psi$ for all $x \in \text{supp}(\psi)$, then $e^{-\frac{1}{(a-1)u_n^{a-1}}} \geq C$ in $\text{supp}(\psi)$, therefore

$$\int_{\Omega} e^{-\frac{1}{(a-1)u_n^{a-1}}} |\nabla T_k(u_n) - \nabla T_k(u)|^2\psi dx \geq C\int_{\Omega} |\nabla T_k(u_n) - \nabla T_k(u)|^2\psi dx.$$

So we conclude that

$$C\int_{\Omega} |\nabla T_k(u_n) - \nabla T_k(u)|^2\psi dx = o(1)$$

and the result follows.

Now it is clear that $\Big(\dfrac{1}{(u_n + \frac{1}{n})^a}\Big)\dfrac{|\nabla u_n|^q}{1 + \frac{1}{n}|\nabla u_n|^q}\psi \to \dfrac{|\nabla u|^q}{u^a}\psi$ a.e. in Ω.

Therefore if $E \subset \Omega$ is a measurable set then

$$\begin{aligned}
&\int_E \Big(\frac{1}{(u_n + \frac{1}{n})^a}\Big)\frac{|\nabla u_n|^q}{1 + \frac{1}{n}|\nabla u_n|^q}\psi dx \\
&= \int_{E\cap\{u_n<k\}} \Big(\frac{1}{(u_n + \frac{1}{n})^a}\Big)\frac{|\nabla T_k(u_n)|^q}{1 + \frac{1}{n}|\nabla T_k(u_n)|^q}\psi dx \\
&+ \int_{E\cap\{u_n\geq k\}} \Big(\frac{1}{(u_n + \frac{1}{n})^a}\Big)\frac{|\nabla u_n|^q}{1 + \frac{1}{n}|\nabla u_n|^q}\psi dx
\end{aligned}$$

The first term in the right hand side obviously converges. We deal with the last term,

$$\int_{E\cap\{u_n\geq k\}}\Big(\frac{1}{(u_n+\frac{1}{n})^a}\Big)\frac{|\nabla u_n|^q}{1+\frac{1}{n}|\nabla u_n|^q}\psi\,dx \leq \int_{u_n\geq k}\frac{|\nabla u_n|^q}{u_n^a}\psi\,dx \leq \frac{1}{k^s}\int_{u_n\geq k}\frac{|\nabla u_n|^q}{u_n^{a-s}}\psi\,dx$$

Using (3.36), choosing $s << 1$, and by Young inequality we obtain that

$$\frac{1}{k^s}\int_{u_n\geq k}\frac{|\nabla u_n|^q}{u_n^{a-s}}\psi\,dx \leq \frac{C}{k^s}.$$

Hence using Vitali lemma we conclude that $\Big(\frac{1}{(u_n+\frac{1}{n})^a}\Big)\frac{|\nabla u_n|^q}{1+\frac{1}{n}|\nabla u_n|^q}\psi \to \frac{|\nabla u|^q}{u^a}\psi$ strongly in $L^1(\Omega)$ and, as a consequence, the existence result follows. □

Remark 3.8 The above existence result is independent of the size of the datum. Notice also that the argument used above to get the local strong convergence of $T_k(u_n)$ is nonlinear in nature. This fact is relevant for the study the nonlinear operators in divergence form.

3.2.4 Some Remarks on Multiplicity if $q = 2$ and $-1 < 2\alpha < 0$

The results in this section extend to the singular framework those obtained in Sect. 1 for increasing nonlinearities. We have the following extreme cases.

(a) $-\Delta u = \beta(u)|\nabla u|^2 + \lambda f$, β a nondecreasing continuous function. Under this hypotheses we find infinitely many solutions of fine energy for suitable f and λ small.
(b) $-\Delta u = \beta(u)|\nabla u|^2 + \lambda f$ with $\beta \geq 0$ continuous and such that

$$\int_0^\infty b(s)ds < \infty,$$

at most there exists a solution. See the details in [95].

We will study here the intermediate situation when $q = 2$ and $2\alpha \geq -1$ The main result is the following.

Theorem 3.9 *Assume that $q = 2$ and $-1 \leq 2\alpha < 0$. Let f be a nonnegative function such that $f \in L^m(\Omega)$, $m > \frac{N}{2}$, then for all $\lambda > 0$, problem (3.2) has infinitely many positive solutions, at least one is bounded, verifying for all $r < \frac{N}{N-1}$,*

$$\int_\Omega e^{\frac{r}{1+2\alpha}u^{1+2\alpha}}|\nabla u|^r dx < \infty \quad \textit{if}\ \alpha > -\frac{1}{2}$$

and

$$\int_\Omega u^r |\nabla u|^r dx < \infty \quad \text{if } \alpha = -\frac{1}{2}.$$

Proof Assume that $-1 \le 2\alpha < 0$. We proceed as follows. Define $v = H(u)$ where H is regular to justify the calculus. If is a solution of (3.21), by a direct calculation we find

$$-\Delta v = -H''(u)|\nabla u|^2 + u^{2\alpha} H'(u)|\nabla u|^2 + H'(u)\lambda f$$

We impose that

$$-H''(u) + u^{2\alpha} H'(u) = 0,$$

that is,

$$H(s) = \begin{cases} \frac{1}{2}s^2 & \text{if } 2\alpha = -1 \\ \displaystyle\int_0^s e^{\frac{t^{2\alpha+1}}{2\alpha+1}} dt & \text{if } 2\alpha > -1. \end{cases}$$

Calling $D(s) \equiv H'(H^{-1}(s))$, we have precisely,

$$D(s) = \begin{cases} \sqrt{2s} & \text{if } 2\alpha = -1 \\ e^{\frac{(H^{-1}(s))^{2\alpha+1}}{2\alpha+1}} & \text{if } -1 < 2\alpha < 0 \end{cases}$$

and v satisfy the problem

$$\begin{cases} -\Delta v = \lambda f D(v) & \text{in } \Omega \\ v = 0 & \text{on } \partial\Omega, \end{cases} \tag{3.41}$$

We claim that $\dfrac{D(s)}{s}$ is a strictly decreasing function in $(0, \infty)$.

The case $2\alpha = -1$ is trivial, so we assume that $2\alpha > -1$. Then

$$\left(\frac{D(s)}{s}\right)' = \frac{sD'(s) - D(s)}{s^2}.$$

A direct computation shows that

$$sD'(s) - D(s) = s(H^{-1}(s))^{2\alpha} - H'(H^{-1}(s)).$$

We set $t = H^{-1}(s)$, then

$$sD'(s) - D(s) = t^{2\alpha} H(t) - H'(t).$$

Since $-1 < 2\alpha < 0$, then $t^{2\alpha} H(t) - H'(t) < 0$ in $(0, \infty)$ and then the result follows.

Then, in particular, $D(s)$ is sublinear in $(0, \infty)$.

Let μ_s be a bounded nonnegative Radon measure which is concentred on a set U of zero capacity and consider v, the unique renormalized solution to problem

$$\begin{cases} -\Delta v = \lambda f D(v) + \mu_s & \text{in } \Omega \\ v = 0 & \text{on } \partial\Omega, \end{cases} \tag{3.42}$$

Notice that the uniqueness of v follows directly by combining the results of [20, 45] and [120].

We can construct v as a limit of $\{v_n\}$ where v_n is the unique solution of the problem

$$\begin{cases} -\Delta v_n = \lambda f D(v_n) + h_n & \text{in } \Omega \\ v_n = 0 & \text{on } \partial\Omega, \end{cases} \tag{3.43}$$

$h_n \in L^\infty(\Omega)$, $||h_n||_{L^1} \le C$ and $h_n \rightharpoonup \mu_s$ in the sense of measures.

Using $1 - \frac{1}{(1+v_n)^\theta}$, $\theta > 0$, as a test function in (3.43) we conclude that

$$\int_\Omega \frac{|\nabla v_n|^2}{(1+v_n)^{1+\theta}} dx \le C.$$

Define $u_n = H^{-1}(v_n)$. By a direct computation we show that

$$-\Delta u_n = u_n^{2\alpha} |\nabla u_n|^2 + \lambda f(x) + \frac{h_n}{D(v_n)}.$$

Therefore we have just to prove that $\dfrac{h_n}{D(v_n)} \to 0$ in the sense of distributions. Since $\dfrac{D(s)}{s} \to 0$ as $s \to \infty$, we need to introduce some modification of the argument used in Theorem 1.35 of Sect. 1 (see too [4]).

If $2\alpha > -1$, then $D(0) = 1$ and $\dfrac{D(s)}{s^\sigma} \to \infty$ as $s \to \infty$ for all $0 < \sigma < 1$.

If $2\alpha = -1$, then $D(s) = \sqrt{2s}$.

Thus in both cases $D(s) \ge C\sqrt{s}$, then $\dfrac{h_n}{D(v_n)} \le C \dfrac{h_n}{\sqrt{v_n}}$.

Let U_ϵ be an open set, $U \subset U_\epsilon$, such that $\text{cap}_{1,2}(U_\epsilon) < \epsilon$. Consider $\phi \in C_0^\infty(\Omega)$, $||\phi||_{W_0^{1,2}} < \epsilon$ and $\phi(x) \ge 1$ if $x \in U_\epsilon$, by using Picone inequality in Lemma 1.3 to

v_n we obtain that

$$\int_\Omega |\nabla\phi|^2 dx \ge \int_\Omega \frac{-\Delta v_n}{v_n}\phi^2 dx \ge \int_\Omega \frac{h_n}{v_n}\phi^2 dx$$

Thus $\int_{U_\varepsilon} \frac{h_n}{v_n} dx \le \varepsilon$. Therefore, using Hölder inequality there result that

$$\int_{U_\varepsilon} \frac{h_n\phi}{D(v_n)} dx \le C\int_{U_\varepsilon} \frac{h_n|\phi|}{\sqrt{v_n}} dx$$

$$\le C\Big(\int_{U_\varepsilon} \frac{h_n}{2v_n} dx\Big)^{\frac12}\Big(\int_{U_\varepsilon} h_n\phi^2 dx\Big)^{\frac12} \le C\varepsilon^{\frac12}.$$

Using now the fact that μ_s is concentrated in $U \subset U_\varepsilon$ and that $D(s) \ge 1$, for every $\phi \in C_0^\infty(\Omega)$ we get

$$\int_{\Omega\setminus U_\varepsilon} |\phi|\frac{h_n}{D(v_n)} dx\, dx \to 0 \text{ for } n \to \infty.$$

That is, combining the above estimates, we conclude that $\frac{h_n}{D(v_n)} \to 0$ in the sense of distributions. Finally, following closely the arguments in the proof of Theorem 1.21, we obtain that $u_n \to u$ in $W_0^{1,2}(\Omega)$ and that u solves

$$\begin{cases} -\Delta u = u^{2\alpha}|\nabla u|^2 + \lambda f(x) & \text{in } \Omega \\ u = 0 & \text{on } \partial\Omega. \end{cases} \tag{3.44}$$

Hence for all μ_s concentrated in a subset of Ω with zero capacity we find a solution. Therefore the multiplicity result follows.

To reach the regularity result on u we use the fact that $v \in W_0^{1,r}(\Omega)$ for all $r < \frac{N}{N-1}$. Finally in the case where $\mu \equiv 0$, then v is bounded and then also u is bounded. □

It is interesting to point out that if $q < 2$, and $-1 \le \alpha q < 0$, using the arguments above we can prove the existence of a positive solution to problem (3.2) under weaker assumptions on f. More precisely we have the following result.

Theorem 3.10 *Assume that $q < 2$ and $-1 \le \alpha q < 0$, recall that $a = -q\alpha$, then*

1. *If $-\alpha \ge \frac12$, for all $f \in L^\sigma(\Omega)$ where $\sigma > \frac{q}{q-a}$, problem* (3.2) *has a distributional solution u such that $u^{1+\frac{\gamma}{q}} \in W_0^{1,q}(\Omega)$ where $\gamma = (q-a)\sigma - q > 0$.*
2. *If $-\alpha < \frac12$, for all $f \in L^m(\Omega)$, $m > \frac{N}{2}$, problem* (3.2) *has a distributional solution u such that $u \in L^\infty(\Omega) \cap W_0^{1,q}(\Omega)$.*

See the detail in [7].

3.2.5 The Case $0 \le q\alpha$

In this section we consider the case $q\alpha \ge 0$. If $\alpha = 0$ there exists a large literature about this problem. If $q = 2$ and $\alpha > 0$ the problem has been studied in [4]. Hence in this section we concentrate the attention in the case $q \in (1, 2)$ and $\alpha > 0$.

Given a $g \in L^1(\Omega)$, $g \gneqq 0$ we define

$$\lambda_1(g) = \inf_{\phi \in W_0^{1,2}(\Omega)} \frac{\displaystyle\int_\Omega |\nabla \phi|^2 dx}{\displaystyle\int_\Omega g\phi^2 dx}.$$

As in Sect. 1, we will assume that

$$(H) \qquad \lambda_1(g) > 0$$

We will use the next existence result whose proof is a simple modification of the result obtained in [85], see also [4].

Theorem 3.11 *Assume that $g \in L^1(\Omega)$ is a nonnegative function such that the hypothesis (H) holds. Then for all $\theta > 0$, for all $c > 0$ and for all $\lambda < \lambda_1$, the problem*

$$\begin{cases} -\Delta w = |\nabla w|^2 + cw^\theta + \lambda g(x) & \text{in } \Omega \\ \quad w = 0 & \text{on } \partial\Omega, \end{cases} \tag{3.45}$$

has at least a positive solution w such that $e^{\frac{\sigma}{2}w} - 1 \in W_0^{1,2}(\Omega)$ for all $\sigma < 1$.

The main existence result of this section is the following.

Theorem 3.12 *Assume that $f \in L^1(\Omega)$ is such that $\lambda_1(f) > 0$. Then there exists $0 < \lambda_* \le \lambda_1(f)$ such that the problem (3.2), with $\alpha > 0$, has a positive minimal solution in $W_0^{1,2}(\Omega)$ for all $\lambda < \lambda_*$.*

Proof Since $q < 2$, then using Young inequality we obtain that

$$s^{q\alpha} t^q \le t^2 + c_1 s^{\frac{2q\alpha}{2-q}}.$$

For simplicity of notation we set $\theta = \frac{2q\alpha}{2-q}$. Since $\lambda_1(f) > 0$ and $\lambda < \lambda_1$, then, from Theorem 3.11, there exists $w \in W_0^{1,2}(\Omega) \cap L^{\theta+1}(\Omega)$, the minimal solution to problem

$$-\Delta w = |\nabla w|^2 + c_1 w^\theta + \lambda f,$$

It is clear that w is a supersolution to problem (3.2). We proceed now by approximation. Let u_n be the minimal solution to problem

$$\begin{cases} -\Delta u_n = T_n^{q\alpha}(u_n)\dfrac{|\nabla u_n|^q}{1+\frac{1}{n}|\nabla u_n|^q} + \lambda f(x) & \text{in } \Omega \\ u_n = 0 & \text{on } \partial\Omega, \end{cases} \tag{3.46}$$

Since w is a supersolution to problem (3.46), then by a simple modification of Lemma 3.1, we obtain that

$$u_n \le u_{n+1} \le w \text{ for all } n.$$

Hence there exist limits

$$u = \lim_{n\to\infty} u_n.$$

By the comparison principle we have that $u_n \le u \le w$ for all n. Let us show that $\{u_n\}$ is bounded in $W_0^{1,2}(\Omega) \cap L^{\theta+1}(\Omega)$.

Since $u_n \le w \in W_0^{1,2}(\Omega) \cap L^{\theta+1}(\Omega)$, then $\{u_n\}$ is bounded in $L^{\theta+1}(\Omega)$.

Using the fact that $-\Delta u_n \ge 0$ we get

$$\int_\Omega |\nabla u_n|^2 dx = \int_\Omega -\Delta u_n u_n dx \le \int_\Omega -\Delta u_n w dx \le \left(\int_\Omega |\nabla u_n|^2 dx\right)^{\frac{1}{2}}\left(\int_\Omega |\nabla w|^2 dx\right)^{\frac{1}{2}}.$$

Thus $\{u_n\}$ is bounded in $W_0^{1,2}(\Omega) \cap L^{\theta+1}(\Omega)$. By the compactness result in [12] we conclude that $u_n \to u$ strongly in $W_0^{1,2}(\Omega)$. Thus

$$T_n^{q\alpha}(u_n)\frac{|\nabla u_n|^q}{1+\frac{1}{n}|\nabla u_n|^q} \to u^{q\alpha}|\nabla u|^q \text{ a.e in } \Omega.$$

By using Vitali's and dominated convergence Theorems and the regularity of w, it follows that

$$T_n^{q\alpha}(u_n)\frac{|\nabla u_n|^q}{1+\frac{1}{n}|\nabla u_n|^q} \to u^{q\alpha}|\nabla u|^q \text{ strongly in } L^1(\Omega).$$

Thus the existence result follows. □

To complete this section we prove a non existence result for λ large.

Theorem 3.13 *There exists λ^* such that if $\lambda > \lambda^*$, problem* (3.2), *has no positive distributional solution.*

Proof Without loss of generality we can assume that $f \in L^\infty(\Omega)$. We argue by contradiction. Assume that for all $\lambda > 0$, problem (3.2) has a distributional solution u. Let v be the solution to problem $-\Delta v = f$ in Ω and $v = 0$ on $\partial\Omega$, then $v \le u$. Thus u satisfies

$$-\Delta u \ge v^{q\alpha}|\nabla u|^q + \lambda f. \tag{3.47}$$

Fixed a sub domain $\Omega_0 \subset\subset \Omega$, by using the strong maximum principle, there exists a positive constant c such that $v(x) \ge c$ for all $x \in \Omega_0$. Let $\phi \in C_0^\infty(\Omega_0)$, using $|\phi|^{q'}$ as a test function in (3.47) it follows that

$$q' \int_{\Omega_0} |\phi|^{q'-1}|\nabla u||\nabla\phi|dx \ge c^{q\alpha} \int_{\Omega_0} |\nabla u|^q|\phi|^{q'} dx + \lambda \int_{\Omega_0} f|\phi|^{q'} dx.$$

Using Young and Hölder inequalities we obtain that

$$C \int_{\Omega_0} |\nabla\phi|^{q'} dx \ge \lambda \int_{\Omega_0} f|\phi|^{q'} dx,$$

then necessarily, $\lambda \le C \inf_{\phi \in C_0^\infty(\Omega_0)} \dfrac{\int_\Omega |\nabla\phi|^{q'} dx}{\int_\Omega f\phi^{q'} dx}$ and this is a contradiction. Hence there exists λ^* such that if $\lambda > \lambda^*$ problem (3.2) has no positive distributional solution. □

3.3 *The Evolution Problem*

In this section we will consider the parabolic problem

$$\begin{cases} u_t - \Delta u^m = |\nabla u|^q + f(x,t), & u \ge 0 \text{ in } \Omega_T \equiv \Omega \times (0,T), \\ \quad u(x,t) = 0 & \text{on } \partial\Omega \times (0,T), \\ \quad u(x,0) = u_0(x), & \text{in } \Omega, \end{cases} \tag{3.48}$$

where $\Omega \subset \mathbb{R}^N$, is a smooth bounded domain, $N \ge 1, m > 0, 1 < q \le 2$, and $f \ge 0$, $u_0 \ge 0$, are in a suitable class of measurable functions. We recall that if $m > 1$, problem (3.48) is a model of growth in a porous medium, see again [21]. We will also consider $1 > m > 0$, that is, the so called *fast diffusion equation*.

As we point out in the introduction of this chapter, performing the change of variable $v = u^m$ problem (3.48) becomes

$$b(v)_t - \Delta v = v^{q(\frac{1}{m}-1)}|\nabla v|^q + f(x,t) \text{ with } b(s) = s^{\frac{1}{m}}.$$

The equation

$$b(v) - \Delta v = \mu$$

usually is known in the literature as *elliptic-parabolic equation*. References for problems related to these equations are [14, 24, 25, 31, 49, 50] and [124] among others.

We will study the *elliptic-parabolic problems* with μ a bounded Radon measure, which is the natural class of data in the application to the analysis of problem (3.48).

The strategy that we follow to study problem (3.48) can be summarized in the following points.

1. We consider approximated problems that kill the degeneration, or the singularity, in the principal part ((PME), or (FDE), respectively) and we truncate the first order term in the right hand side. With respect to these approximated problems, the existence of a solution follows using the well known results obtained in [36]. Here the natural setting is to find a *weak solution* (which is formulated for the corresponding elliptic-parabolic equation).
2. We obtain uniform estimates of the solution of the approximated problems in such a way that the first order part in the second member is uniformly bounded in $L^1(\Omega_T)$.
3. The previous step motivates the study of a problem with measure data. To have more flexibility in the calculation we formulate the problem as an elliptic-parabolic equation and look for a *reachable solution* in the sense given in the next section. We will skip the very technical proof of the almost everywhere convergence of the gradients of the solutions of the approximated problems, sending to the reader to the reference [9].
4. The final step is to use the uniform estimates and the a.e. convergence of the gradients to prove that, up to a subsequence, the second members of the approximated problems of (3.48), converge strongly in $L^1_{loc}(\Omega_T)$. That is, we find a *distributional solution*.

3.3.1 Review to Some Results for an *Elliptic-Parabolic* Problem with Measure Data: Applications to (PME) and (FDE) Equations

We will consider the problem

$$\begin{cases} \big(b(v)\big)_t - \Delta\, v = \mu & \text{in } \Omega \times (0, T)\,, \\ v(x,t) = 0 & \text{on } \partial\Omega \times (0, T)\,, \\ b(v(x,0)) = b(v_0(x)) & \text{in } \Omega\,, \end{cases} \tag{3.49}$$

where $b : \mathbb{R} \to \mathbb{R}$ is continuous strictly increasing function such that $b(0) = 0$, $b(v_0) \in L^1(\Omega)$ and μ is a Radon measure whose total variation is finite in Ω_T. We will assume the following hypotheses on b:

$$
(B)\begin{cases}
(B1) \text{ There exists } a_1 > 0 \text{ such that } b(s) \geq Cs^{a_1} \text{ for } s >> 1,\\
(B2) \text{ There exists } a_2 < 1 \text{ such that } |b'(s)| \leq \dfrac{1}{s^{a_2}} \text{ for } s << 1,\\
(B3) \text{ There exits } a_3 \in (\dfrac{N-1}{N}, 1) \text{ such that for some } \varepsilon > 0:\\
\text{Either } b' \in C([0,\infty)) \text{ and } |b'(s)|\, b^{2a_3-1}(s) \leq s^{\frac{N+2a_1}{N}-\varepsilon} \text{ as } s \to \infty\\
\text{or } |b'(s)| \leq b^{2-2a_3-\varepsilon}(s) \text{ as } s \to \infty.
\end{cases}
$$

Examples The following examples of b will be considered in this section.

1. $b(s) = s^{\sigma}$ if $s \geq 0$, for some $\sigma > \dfrac{(N-2)_+}{N}$.
2. $b(s) = \dfrac{1}{m}\displaystyle\int_0^s (\Lambda^{-1}(\sigma))^{\frac{1}{m}-1} d\sigma = \dfrac{1}{m}\int_0^{\Lambda^{-1}(s)} \sigma^{\frac{1}{m}-1}\Lambda'(\sigma)d\sigma$, where

$$\Lambda(s) = \frac{4}{5}s^{\frac{5}{4}} \textit{ if } m = 2, \quad \Lambda(s) = \int_0^s e^{\frac{t^{\frac{2-m}{m}}}{m(2-m)}} dt \textit{ if } 0 < m < 2.$$

3. $b_n(s) := b(s + b(\frac{1}{n})) - b(\frac{1}{n})$, $b(s) = s^{\sigma}$ and $\sigma > \dfrac{(N-2)_+}{N}$. In this case we are able to show that the estimates obtained for approximated problems are uniform in $n \geq 1$. A similar observation can be done for the truncation of Λ.

Definition 3.14 Assume that $\mu \in L^{\infty}(\Omega_T)$ and $b(v_0) \in L^{\infty}(\Omega)$. We say that v is a weak solution to (3.49) if 1) $v \in L^2((0,T));\, W_0^{1,2}(\Omega)) \cap L^{\infty}(\Omega_T)$; 2) The function $b(v) \in C((0,T); L^q(\Omega))$ for all $q < \infty$; 3) $(b(v))_t \in L^2((0,T); W^{-1,2}(\Omega))$ and for every $\phi \in L^2((0,T); W_0^{1,2}(\Omega))$ the following identity holds,

$$\int_0^T \langle b(v)_t, \phi\rangle + \iint_{\Omega_T} \nabla v \cdot \nabla\phi = \iint_{\Omega_T} \mu\,\phi\,. \tag{3.50}$$

For bounded data, the following result is well known.

Theorem 3.15 *Assuming $\mu \in L^{\infty}(\Omega_T)$ and $b(v_0) \in L^{\infty}(\Omega)$, there exists a unique weak solution to problem* (3.49) *in the sense of Definition 3.14.*

The proof of Theorem 3.15 can be found in [14] and [50].

Reachable Solutions Since we are considering problems with general data, in particular with measure data, we need to precise the sense in which the solution is defined. For elliptic equations the notion of *reachable solutions* was introduced in [60]. We refer to [54] for the parabolic equation. See also [6] for some particular cases. If $\mu \in L^1(\Omega_T)$, the renormalized solution is studied in [31].

It is worthy to point out that (PME) and (FDE) with measure data has been recently studied by T. Lukkari in [105] and [106] respectively. The approach that we do here is different. We consider the associated *elliptic-parabolic equation* and the corresponding approximated problems in order to prove the almost everywhere convergence of the gradients that we need to apply the results to equations with a gradient term.

Here after we assume that b verifies hypothesis (B) and we omit an explicit reference to this fact.

In our framework, we consider the following definition of *reachable solution*.

Definition 3.16 Assume that μ is a Radon measure whose total variation is finite in Ω_T, b verifies hypothesis (B) and $b(v_0) \in L^1(\Omega)$.

We say that v is a reachable solution to (3.49) if

1. $T_k(v) \in L^2((0,T); W_0^{1,2}(\Omega))$ for all $k > 0$.
2. For all $t > 0$ there exist both one-side limits $\lim\limits_{\tau \to t^{\pm}} b(v(\cdot,\tau))$ weakly-* in the sense of measures.
3. $b(v(\cdot,t)) \to b(v_0(\cdot))$ weakly-* in the sense of measures as $t \to 0$.
4. There exist three sequences $\{v_n\}_n$ in $L^2((0,T); W_0^{1,2}(\Omega))$, $\{h_n\}_n$ in $L^\infty(\Omega_T)$ and $\{g_n\}_n$ in $L^\infty(\Omega)$ such that if v_n is the weak solution to problem

$$\begin{cases} \big(b(v_n)\big)_t - \Delta\, v_n = h_n & \text{in } \Omega \times (0,T)\,, \\ v_n(x,t) = 0 & \text{on } \partial\Omega \times (0,T)\,, \\ v_n(x,0) = b^{-1}\big(g_n(x)\big) & \text{in } \Omega\,, \end{cases} \tag{3.51}$$

then

(a) $g_n \to b(v_0)$ in $L^1(\Omega)$.
(b) $h_n \overset{*}{\rightharpoonup} \mu$ as measures.
(c) $\nabla v_n \to \nabla v$ strongly in $L^\sigma(\Omega_T)$ for $1 \le \sigma < \dfrac{N+2a_1}{N+a_1}$.
(d) The sequence $\{b(v_n)\}_n$ is bounded in $L^\infty((0,T); L^1(\Omega))$ and $b(v_n) \to b(v)$ strongly in $L^1(\Omega_T)$.

In order to solve the problem (3.49) we consider the approximating problems,

$$\begin{cases} b(v_n)_t - \Delta v_n = h_n, & (x,t) \in \Omega_T, \\ v_n(x,t) = 0, & (x,t) \in \partial\Omega \times (0,T), \\ v_n(x,0) = b^{-1}(g_n(x)), & x \in \Omega, \end{cases} \tag{3.52}$$

where $g_n \to b(v_0)$ strongly in $L^1(\Omega)$ and $h_n \to \mu$ in the weak-$*$ sense in Ω_T. The existence of weak solutions to these problems follows from Theorem 3.15. Moreover, by taking suitable test functions we find the following a priori estimates.

Proposition 3.17 *Let $\{v_n\}_n$ be a sequence of solutions of the approximate problems* (3.52). *Then*

1. *For each $0 < \beta < \frac{1}{2}$, the sequence $\{(|v_n| + 1)^\beta - 1\}_n$ is bounded in $L^2(0, T; W_0^{1,2}(\Omega))$*
2. *The sequence $\{b(v_n)\}_n$ is bounded in the space $L^\infty(0, T; L^1(\Omega))$ and $\{(b(v_n))_t\}_n$ is bounded in $L^1(\Omega_T) + L^\sigma(0, T; W^{-1,\sigma}(\Omega))$, for some $\sigma > 1$.*
3. *For all $\alpha > 0$,*

$$\iint_{\Omega_T} \frac{|\nabla v_n|^2}{(1 + |v_n|)^{\alpha+1}} \leq C \tag{3.53}$$

Furthermore, the sequence $\{|\nabla v_n|\}_n$ is bounded in the Marcinkiewicz space $M^q(\Omega_T)$, for $q = \frac{(N+2a_1)}{N+a_1}$ and $\{v_n\}_n$ is bounded in the space $M^\sigma(\Omega_T)$, where $\sigma = \frac{N+2a_1}{N}$.

See the details of the proof in [8]. Moreover, and this is new with respect to the results in [105] and [106], we obtain the almost everywhere convergence of the gradients. Precisely we get the following result.

Proposition 3.18 *Consider $\{v_n\}_n$, the solution of the approximated problems* (3.52). *Then, up to subsequence,*

$$\nabla T_k(v_n) \to \nabla T_k(v) \qquad \textit{almost everywhere in } \Omega_T\,. \tag{3.54}$$

As a consequence, $\nabla v_n \to \nabla v$ almost everywhere in Ω_T.

Hence we can formulate the following Theorem.

Theorem 3.19 *Let b be a function verifying the hypotheses (B). Let μ be a finite Radon measure, and consider $\{v_n\}_n$ is a sequence of solutions to* (3.52). *Then there exists a measurable function v which is a reachable solution to the problem* (3.49), *namely,*

1. *$\nabla v_n \to \nabla v$ strongly in $L^\sigma(\Omega_T)$ for all $1 \leq \sigma < \frac{N+2a_1}{N+a_1}$.*
2. *For all $\sigma < 2$,*

$$\nabla T_k(v_n) \to \nabla T_k(v) \textit{ strongly in } L^\sigma(\Omega_T). \tag{3.55}$$

3. *For every $\Phi \in C^\infty(\overline{\Omega}_T)$, such that $\Phi(\cdot, t) \in C_0(\Omega)$ for all $t \in (0, T)$ and $\Phi(x, T) = 0$ for all $x \in \Omega$, the following identity holds*

$$-\int_\Omega b(v_0(x))\Phi(x, 0)\,dx - \iint_{\Omega_T} b(v)\Phi_t + \iint_{\Omega_T} \nabla v \cdot \nabla \Phi = \iint_{\Omega_T} \Phi\, d\mu\,. \tag{3.56}$$

Application to the Porous Media and Fast Diffusion Equations with a Radon Measure The results in the above subsection allow us to consider the problem

$$\begin{cases} u_t - \Delta u^m = \mu & \text{in } \Omega \times (0,T)\,, \\ u(x,t) = 0 & \text{on } \partial\Omega \times (0,T)\,, \\ u(x,0) = u_0(x) & \text{in } \Omega\,, \end{cases} \tag{3.57}$$

with $m > \dfrac{(N-2)_+}{N}$, $u_0 \in L^1(\Omega)$ and μ is a Radon measure whose total variation is finite in Ω_T.

Remark 3.20 If $0 < m \leq \dfrac{(N-2)_+}{N}$, among others, the difficulty is to show the strong convergence of the sequence $\{u_n\}_n$ in $L^1(\Omega_T)$. This can be proved by assuming additional hypotheses on the source term μ. See [9] and too [106]. Once proving this strong convergence, the result of the Theorem 3.21 holds with the same conclusions.

We can use directly the result of Theorem 3.19 to get the existence of a reachable solution to problem (3.57). However we will use an equivalent approach that will be useful in the following sections to analyze the truncated problems in (*PME*) and (*FDE*) with a gradient term. In the case of the porous media equation, i.e., $m > 1$, the existence results are obtained in [105] by using some result in [91], however our approach is different and follows using the *elliptic-parabolic*.

The (FDE) with $1 > m > \dfrac{(N-2)_+}{N}$ was obtained independently in [9] and [106] and with some different arguments. Our approach use the *elliptic-parabolic* framework discussed above that allow us to prove the a.e convergence of the gradients of the truncated problems to the gradient of the solution of problem (3.57), that is a key step to apply the results to (PME) and (FDE). We will consider the approximated form

$$\begin{cases} u_{nt} - \operatorname{div}(m(u_n + \frac{1}{n})^{m-1}\nabla u_n) = h_n & \text{in } \Omega \times (0,T)\,, \\ u_n(x,t) = 0 & \text{on } \partial\Omega \times (0,T)\,, \\ u_n(x,0) = T_n(u_0) & \text{in } \Omega. \end{cases} \tag{3.58}$$

The main goal is to show compactness results for the sequences $\{|\nabla u_n|\}_n$ and $\{T_k(u_n)\}_n$ for

$$\frac{(N-2)_+}{N} < m < 1.$$

Define $v_n \equiv (u_n + \frac{1}{n})^m - (\frac{1}{n})^m$, then v_n solves

$$\begin{cases} (b_n(v_n))_t - \Delta v_n = h_n, & (x,t) \in \Omega_T, \\ v_n(x,t) = 0, & (x,t) \in \partial\Omega \times (0,T), \\ v_n(x,0) = \varphi^{-1}(T_n(u_0(x))), \; x \in \Omega, & \end{cases} \tag{3.59}$$

where $b_n(s) = (s + (\frac{1}{n})^m)^{\frac{1}{m}} - \frac{1}{n}, n \ge 1$.

Applying Theorem 3.19 we obtain the following result.

Theorem 3.21 *Consider v_n and $u_n = b(v_n)$, the solutions to (3.59) and (3.58) respectively. Then, there exists a measurable function u such that $u^m \in L^r(0,T; W_0^{1,r}(\Omega))$ for all $r < 1 + \frac{1}{Nm+1}$, and, up to a subsequence,*

1. *$\nabla u_n \to \nabla u$ e.a in Ω_T and then $\nabla v_n \to \nabla v$ e.a in Ω_T where $v = u^m$.*
2. *$T_k(v_n) \to T_k(v)$ strongly in $L^\sigma(0,T; W_0^{1,\sigma}(\Omega))$ for all $k > 0$ and for all $\sigma < 2$.*

For a detailed proof the reader can see [9].

3.3.2 The Porous Medium Equations with a Gradient Term

We arrive to one of the main proposed question, that is, to prove existence of solution to Problem (3.48). The proof of the existence result is a consequence of the following steps.

1. We prove some a priori estimates that allow us to show that the right hand side of the truncated problems converge weak-* to a Radon measure. Precisely we consider the approximated problems

$$\begin{cases} u_{nt} - \operatorname{div}(m(u_n + \frac{1}{n})^{m-1}\nabla u_n) = \dfrac{|\nabla u_n|^q}{|\nabla u_n|^q + \frac{1}{n}} + f_n & \text{in } \Omega_T, \\ u_n(x,t) = 0 & \text{on } \partial\Omega \times (0,T), \\ u_n(x,0) = u_{0n}(x)) & \text{if } x \in \Omega. \end{cases} \tag{3.60}$$

   Notice that the existence and the boundedness of u_n follow using the results in [36]. We are able to obtain a uniform bound of the L^1-norm of the righthand side. Then up to a subsequence such second terms weakly-* converge to a Radon measure.
2. We transform in a natural way the problem to an *elliptic-parabolic* problem. By using the results of Theorem 3.21 and some compactness arguments, we identify the measure limit as the second member of the Problem (3.48).

As in the stationary problem we have a different behavior of the problem according the values of m.

The Case $1 < m \leq 2$

Consider the problem

$$\begin{cases} u_t - \Delta u^m = |\nabla u|^q + f(x,t) & \text{in } \Omega_T \equiv \Omega \times (0,T), \\ u(x,t) \geq 0 & \text{in } \Omega_T, \\ u(x,t) = 0 & \text{on } \partial\Omega \times (0,T), \\ u(x,0) = u_0(x) & \text{if } x \in \Omega, \end{cases} \tag{3.61}$$

where $m > 1$, $q \leq 2$, $\Omega \subset \mathbb{R}^N$ a bounded domain, f and u_0 nonnegative functions under suitable hypotheses given below.

We will use as starting point the results in [36] for bounded data, $f \in L^\infty(\Omega_T)$ and $u_0 \in L^\infty(\Omega)$. Since $1 < m \leq 2$ and $1 \leq q \leq 2$ we will be able to obtain an a priori estimates in the framework of [105]. Notice that in [91], these estimates are used to analyze the behavior of *viscosity supersolution*, to the porous medium equation. See [91] and [105] for more details concerning to this framework. More precisely we have the next theorem.

Theorem 3.22 *Assume that* $1 < m \leq 2$ *and* $q \leq 2$, *then*

1. *If* $q'(m-1) > 2$, $u_0 \in L^{1+\theta}(\Omega)$ *and* $f \in L^{1+\frac{2\theta}{mN}}(0,T; L^{\frac{(\theta+m)N}{mN+2\theta}}(\Omega))$ *where* $\theta \geq 2-m$, *then problem* (3.61) *has a distributional solution.*
2. *If* $q'(m-1) \leq 2$
   (a) *If* $q < m$, *problem* (3.61) *has a solution for all* f, u_0 *as in the first case.*
   (b) *If* $m \leq q \leq 2$, *then problem* (3.61) *has a solution if* $e^{\alpha u_0} \in L^1(\Omega)$ *for some* $\alpha > 0$ *and* $f \in L^r(0,T; L^s(\Omega))$ *where* $1 < r < \infty$, $s > \frac{N}{2}$ *and* $\frac{1}{r} + \frac{N}{2s} = 1$.

The Case $m > 2$: L^1 *Data*

In the elliptic case if $q(\frac{1}{m} - 1) < -1$, then existence result holds for all L^1 data, without any restriction on its size, see [7]. Notice that, since $q \leq 2$, then the above condition implies that $m > 2$, however, in this section we assume that $m > 2$ and $q \leq 2$, without any other restriction. In particular, our result can be seen as a slight improvement of the result obtained in the elliptic case. The key is to prove some a priori estimates that allow us to show that the problem (3.61) has a distributional solution for all $f \in L^1(\Omega_T)$ and $u_0 \in L^1(\Omega)$. The main existence result in this section is the following.

Theorem 3.23 *Let* f, u_0 *be such that* $f \in L^1(\Omega_T)$ *and* $u_0 \in L^1(\Omega)$. *Assume* $1 < q \leq 2$ *and* $m > 2$, *then problem* (3.61) *has a distributional solution* u *such that* $|\nabla u^m| \in L^\sigma_{loc}(\Omega_T)$ *for all* $1 \leq \sigma < 1 + \frac{1}{Nm+1}$.

We skip the details and refer to the reader to [9].

3.3.3 The Fast Diffusion Equation with a Gradient Term

We consider the case $0 < m < 1$, usually called *fast diffusion equation* in the literature.

We will prove the following existence result.

Theorem 3.24 *Assume that* $0 < m < 1$, $q \leq 2$ *and*

1. $f \in L^r(0,T; L^s(\Omega))$ *where* $1 < r < \infty$, $s > \frac{N}{2}$ *with* $\frac{1}{r} + \frac{N}{2s} = 1$
2. $e^{\alpha u_0^{2-m}} \in L^1(\Omega)$ *where,*
   (*a*) *either* $\alpha > 0$ *is any positive constant if* $q < 2$
   (*b*) *or* $\alpha m(2-m) > 2$ *if* $q = 2$.

Then problem (3.61) *has a distributional solution.*

The regularity required to the source term follows the same motivation as in the stationary problem.

For details of the proof, see [9].

3.3.4 Further Results

We will state some multiplicity results that follows parallel to those in the heat equation and those in the elliptic problem. Also we will give some properties of qualitative behavior of the solutions depending on m in some particular cases. We will follow the references [8] and [9].

Multiplicity Consider the problem

$$\begin{cases} u_t - \Delta u^m = |\nabla u|^2 + f(x,t) & \text{in } \Omega_T \equiv \Omega \times (0,T), \\ u(x,t) \geq 0 & \text{in } \Omega_T, \\ u(x,t) = 0 & \text{on } \partial\Omega \times (0,T), \\ u(x,0) = u_0(x) & \text{if } x \in \Omega, \end{cases} \tag{3.62}$$

where $\Omega \subset \mathbb{R}^N$ is a smooth bounded domain, $0 < m \leq 2$, f and u_0 are nonnegative functions under suitable hypotheses that we will precise below. f $m \neq 1$, by setting $v = H(u^m)$, where H is given by

$$H(s) = \begin{cases} \dfrac{4}{5} s^{\frac{5}{4}} & \text{if } m = 2 \\[2ex] \displaystyle\int_0^s e^{\frac{t^{\frac{2-m}{m}}}{m(2-m)}}\, dt & \text{if } 0 < m < 2, \end{cases} \tag{3.63}$$

we are able to transform problem (3.62) to an *elliptic-parabolic* problem of the form

$$\begin{cases} \big(b(v)\big)_t - \Delta v = f(x,t)D(v), & \text{in } \Omega \times (0,T), \\ v(x,t) = 0 & \text{on } \partial\Omega \times (0,T), \\ b(v(x,0)) = b(v_0(x)) & vin\Omega, \end{cases} \tag{3.64}$$

where D and b are defined by

$$D(s) \equiv H'(H^{-1}(s)) \text{ and } b(s) = \frac{1}{m}\int_0^s (H^{-1}(\sigma))^{\frac{1}{m}-1} d\sigma, \tag{3.65}$$

then

$$D(s) = e^{\frac{(H^{-1}(s))^{\frac{2-m}{m}}}{m(2-m)}} \text{ and } b(s) = \frac{1}{m}\int_0^{H^{-1}(s)} \sigma^{\frac{1}{m}-1} H'(\sigma) d\sigma.$$

The main result on multiplicity is the following.

Theorem 3.25 *Assume that* $0 < m \le 2$. *Let* f *be a nonnegative bounded function and assume that* $u_0 \in L^\infty$, *then problem* (3.62) *has infinitely many positive distributional solutions, verifying* $|\nabla H(u^m)| \in L^\sigma(\Omega_T)$ *for all* $1 \le \sigma < \sigma_0$ *where* H *is defined in* (3.63) *and*

$$\sigma_0 = \begin{cases} 1 + \frac{1}{N+1} & \text{if } m < 2, \\ 1 + \frac{3}{5N+3} & \text{if } m = 2. \end{cases} \tag{3.66}$$

Sketch of the Proof Assume that μ_s is a nonnegative Radon measure singular respect to the parabolic capacity in Q.

Define $v_0 = H(u_0^m)$, and let $\{h_n\}_n$ and $\{g_n\}_n$ be such that $g_n \to b(v_0)$ in $L^1(\Omega)$, $h_n \overset{*}{\rightharpoonup} \mu_s$ as measures. Consider v_n, the solution to the approximate problem

$$\begin{cases} \big(b(v_n)\big)_t - \Delta v_n = f(x,t)\,D(v_n) + h_n & \text{in } \Omega \times (0,T), \\ v_n(x,t) = 0 & \text{on } \partial\Omega \times (0,T), \\ b(v_n(x,0)) = g_n(x) & \text{in } \Omega, \end{cases} \tag{3.67}$$

where b is defined in (3.65).

From the result of Theorem (3.19), there exists a measurable function v such that $v \in L^\sigma(0,T; W_0^{1,\sigma}(\Omega))$ for all $1 \le \sigma < \sigma_0$ defined in (3.66), $T_k(v_n) \to T_k(v)$ strongly in $L^2(0,T; W_0^{1,2}(\Omega))$, $fD(v_n) \to fD(v)$, $b(v_n) \to b(v)$ strongly in $L^1(\Omega_T)$

and v is a reachable solution to the following problem

$$\begin{cases} \big(b(v)\big)_t - \Delta v = f(x,t)\, D(v) + \mu_s & \text{in } \Omega \times (0,T)\,, \\ v(x,t) = 0 & \text{on } \partial\Omega \times (0,T)\,, \\ b(v(x,0)) = b(v_0(x)) & \text{in } \Omega\,. \end{cases} \tag{3.68}$$

We set $u_n = (H^{-1}(v_n))^{\frac{1}{m}}$, then u_n solves

$$(u_n)_t - \Delta u_n^m = |\nabla u_n|^2 + f + \frac{h_n}{D(v_n)} \text{ in } \mathcal{D}'(Q)\,. \tag{3.69}$$

As in the multiplicity result above one can check that,

$$\frac{h_n}{D(v_n)} \to 0 \text{ in } \mathcal{D}'(\Omega_T). \tag{3.70}$$

That is we reach, as a byproduct, the *wild nonuniqueness result* in Theorem 3.25.
The details can be seen in [8].

Some Examples of Finite Time Extinction Assume that $f \equiv 0$ and $q = 2$. Consider the problem

$$\begin{cases} u_t - \Delta u^m = |\nabla u|^2 & \text{in } \Omega_T, \\ u(x,t) = 0 & \text{on } \partial\Omega \times (0,T), \\ u(x,0) = u_0(x) & \text{if } x \in \Omega, \end{cases} \tag{3.71}$$

where $0 < m < 1$.
We precise the meaning of *regular* solutions to (3.71).

Definition 3.26 Let $\Lambda(s) = \displaystyle\int_0^s e^{\frac{t^{\frac{2-m}{m}}}{m(2-m)}}\, dt$ and define

$$\beta(s) = \frac{1}{m} \int_0^s (\Lambda^{-1}(\sigma))^{\frac{1}{m}-1} d\sigma, \tag{3.72}$$

we say that u is a regular solution to problem (3.71) in Ω_T if

$$v \equiv \Lambda(u^m) \in L^2((0,T); W_0^{1,2}(\Omega)) \cap \mathcal{C}([0,T]; L^2(\Omega)),\ \beta(v)_t \in L^2((0,T); W^{-1,2}(\Omega))$$

and for all $\phi \in L^2((0,T); W_0^{1,2}(\Omega))$ we have

$$\int_0^T \langle (\beta(v))_t, \phi \rangle + \int_0^T \int_\Omega \nabla v \cdot \nabla \phi = 0. \tag{3.73}$$

Notice that since $q = 2$ we use a change of variable as in the heat equation studied in this chapter.

It is clear that the existence of a regular solution follows using Theorem 3.24 for $q = 2, f = 0$ and the regularity of the initial datum u_0. Now we are able to state the next result.

Theorem 3.27 *Assume that* $0 < m < 1$*. If* u *is the regular solution of problem* (3.71) *in the sense of Definition 3.26, then there exists a positive, finite time* t_0*, depending on* N*, and* u_0 *such that* $u(x,t) \equiv 0$ *for* $t > t_0$*.*

Proof To get the desired result we have just to show that $v(x,t) \equiv 0$ for $t > t_0$. It is clear that v solves

$$\begin{cases} \big(\beta(v)\big)_t - \Delta v = 0 & \text{in } \Omega \times (0,T)\,, \\ v(x,t) = 0 & \text{on } \partial\Omega \times (0,T)\,, \\ v(x,0) = v_0(x) & \text{in } \Omega\,, \end{cases} \tag{3.74}$$

with $v_0 \in L^2(\Omega)$. Using v^θ, where $\theta > 0$ to be chosen later, as a test function in (3.74), there result that

$$\frac{d}{dt}\int_\Omega \Psi(v(x,t))\,dx + c(\theta)\int_\Omega |\nabla v^{\frac{\theta+1}{2}}|\,dx = 0\,, \tag{3.75}$$

where

$$\Psi(s) = \int_0^s s^\theta (\Lambda^{-1}(\sigma))^{\frac{1}{m}-1}\,d\sigma\,.$$

Since

$$\lim_{s\to\infty} \frac{\Lambda^{-1}(s)}{s^\varepsilon} = 0 \text{ for all } \varepsilon > 0$$

it follows that

$$\Psi(s) \le c(\varepsilon) s^{\theta+1+\varepsilon(\frac{1}{m}-1)} \qquad \text{for every } s \ge 0.$$

Fixed θ such that $\theta + 1 + \varepsilon(\frac{1}{m} - 1) = 1 + \theta_0$, then using Sobolev's and Hölder's inequalities,

$$\int_\Omega |\nabla v^{\frac{\theta+1}{2}}|^2\,dx \ge c_1(N,\theta)\Big[\int_\Omega (v^{\frac{\theta+1}{2}})^{2^*}\,dx\Big]^{2/2^*} \ge c_2(N,\theta,|\Omega|)\Big[\int_\Omega (v^{a(\theta+1)})\,dx\Big]^{1/a}$$

where $1 < a < \frac{2^*}{2}$ is chosen such that $a(\theta+1) = \theta + 1 + \varepsilon(\frac{1}{m} - 1)$. Hence it follows that

$$\int_\Omega |\nabla v^{\frac{\theta+1}{2}}|^2\,dx \ge c(N,\theta,|\Omega|)\Big[\int_\Omega \Psi(v(x,t))\,dx\Big]^{1/a}.$$

Define

$$\xi(t) = \int_\Omega \Psi(v(x,t))\,dx\,,$$

then

$$\frac{\xi'(t)}{\xi(t)^{1/a}} \le -c_4 < 0\,.$$

Notice that by the assumption on v_0 we reach that $\xi(0) < \infty$. Integrating in t, one obtains

$$\frac{a}{a-1}\big(\xi(t)^{\frac{a-1}{a}} - \xi(0)^{\frac{a-1}{a}}\big) \le -c_4 t\,.$$

Thus, as long as $\xi(t) > 0$, one has

$$\xi(t)^{\frac{a-1}{a}} \le \xi(0)^{\frac{a-1}{a}} - c_4\frac{a-1}{a}t\,.$$

Therefore, $\xi(t) \equiv 0$ for t large enough. □

An Example of Finite Speed of Propagation If $m = q = 2$ and $f = 0$, we can prove that the solution to problem (3.61) has the *finite speed propagation property*. This follows by setting $w = \frac{2}{3}\left(\frac{4}{5}\right)^{\frac{2}{5}} u^{\frac{5}{2}}$, then w solves

$$\begin{cases} w_t - \frac{4}{5}\left(\frac{3}{2}\right)^{\frac{5}{3}} \Delta\, w^{\frac{5}{3}} = 0 & \text{in } \Omega_T\,, \\ w(x,t) = 0 & \text{on } \partial\Omega \times (0,T)\,, \\ w(x,0) = \frac{2}{3}\left(\frac{4}{5}\right)^{\frac{2}{5}} u_0^{\frac{5}{2}}(x) & \text{in } \Omega\,. \end{cases} \tag{3.76}$$

If $u_0 \in L^\infty(\Omega)$ has a compact support, by using a convenient Barenblatt self-similar super-solution (see [145], for instance) we obtain the *finite speed of propagation property*. The inverse change of variable allow us to conclude the same result for problem

$$\begin{cases} u_t - \Delta\, u^2 = |\nabla u|^2 & \text{in } \Omega_T\,, \\ u(x,t) = 0 & \text{on } \partial\Omega \times (0,T)\,, \\ u(x,0) = u_0(x) & \text{in } \Omega\,. \end{cases} \tag{3.77}$$

An interesting question in to study the relation between q and m in order to have finite seed of propagation with the growth term. A fist step is try to prove the behavior in the example above without a change of variable.

4 A Fourth Order Model Appearing in Epitaxial Crystal Growth

4.1 Introduction

In this chapter we introduce a variational model to describe epitaxial growth and prove some related mathematical results. There are several remarks to do. First the equations that appear are of fourth order, therefore in general we have not the maximum principle and comparison results. In second place the variational formulation of the problem seems to be deeply sensitive to the boundary condition. And, finally, the functional framework needs of some results involving the compensated-compactness theory (see [112, 141] and [142]) that improve the regularity of the determinant of the differential of vector fields in $\mathbb{R}^N$.

The contents of this chapter are some results in the articles [66, 67] and [72].

4.1.1 About the Model

I would like to thank to my friend Carlos Escudero, in particular, for his help shearing with me the ideas and the motivation of the physical model.

Epitaxial growth is characterized by the deposition of new material on existing layers of the same material under high vacuum conditions. This technique is used, for instance, in the semiconductor industry for the growth of thin films. The crystals grown may be composed of a pure chemical element like silicon or germanium, or may either be an alloy like gallium arsenide or indium phosphide.

In case of molecular beam epitaxy the deposition takes place at a very slow rate and almost atom by atom. The goal in most situations of thin film growth is growing an ordered crystal structure with flat surface. But in epitaxial growth is quite usual finding a mounded structure generated along the surface evolution. The actual origin of this mounded structure is to a large extend unknown, although some mechanisms (like energy barriers) have already been proposed. Attempting to perform ab initio quantum mechanical calculations in this system is computationally too demanding, what opens the way to the development of simplified models: these can be of a discrete probabilistic nature or have the form of a differential equation.

Discrete models usually represent adatoms (the atoms deposited on the surfaces) as occupying lattice sites. They are placed randomly at one such site and then they are allowed to move according to some rules which characterize the different models. One widespread diffusion rule is allowing the displacement of the deposited adatom if this movement increases the coordination number, i.e., the number of first neighbors which are occupied by an adatom. The physics underlying the definition of this rule is the following: the adatom random dispersion minimizes the system chemical potential as the number of chemical bonds among adatoms and the crystalized structure increases.

A different modeling possibility is using differential equations. For instance, this sort of diffusion which increases the coordination number of the random walkers (which model the adatom) is mathematically represented by a heat equation in which the Laplacian is substituted by minus the bilaplacian. In this chapter we focus our approach on modeling by means of differential equations and on the analysis of such models.

A fundamental hypothesis to obtain the mathematical description is that the interface, which describes the height of the growing interface in the spatial point (x, y) at time t, is given by the graph of a function $h = h(x, y, t)$. Although this theoretical framework can be extended to any spatial dimension N, we will concentrate here on the physical situation $N = 2$. A basic assumption is the no overhang approximation, which corresponds with the possibility of parameterizing the interface as a Monge patch.

The macroscopic description of the growing interface is given by a stochastic partial differential equation (SPDE) which is usually postulated using phenomenological arguments. Examples of such theories are given by the well known SPDEs named after Kardar et al. [89] that we study in the previous chapters, models by Edwards and Wilkinson and by Mullins and Herring in [18], and the models by Villain, Lai, and Das Sarma [98].

Herein we will consider a variational formulation of the surface growth equation. In order to proceed with our derivation, we will assume that the height function obeys a gradient flow equation

$$\frac{\partial h}{\partial t} = -\frac{\delta \mathcal{J}}{\delta h} + \xi(x, y, t), \tag{4.1}$$

where we have added the noise term $\xi(x, y, t)$. Here we have:

(I) The functional $\mathcal{J}$ denotes a potential that is pursued to be minimized during the temporal evolution of h.

This potential describes the microscopic properties of the interface and of the adatom interactions and, at large enough scales, we assume that it can be expressed as a function of the surface mean curvature only

$$\mathcal{J} = \int f(H) \sqrt{g}\, dx\, dy, \tag{4.2}$$

where H denotes the mean curvature, g the determinant of the surface metric tensor, and f is an unknown function of H.

(II) We will further assume that this function can be expanded in a power series

$$f(H) = K_0 + K_1 H + \frac{K_2}{2} H^2 + \cdots, \tag{4.3}$$

of which only the zeroth, first, and second order terms will be of relevance at large scales.

The result of the minimization of the potential (4.2) leads to the SPDE

$$\partial_t h = \mu \nabla^2 h + \lambda \left[(\partial_{xx} h)(\partial_{yy} h) - (\partial_{xy} h)^2 \right] - \nu \nabla^4 h + \xi(x, y, t), \tag{4.4}$$

to leading order in the small gradient expansion, which assumes $|\nabla h| \ll 1$, and for suitable constants μ, λ and ν.

The term proportional to μ is common diffusion. In this context it models the random dispersion of the adatoms, which try to move to lower elevation locations. In this sense, this term is a consequence of the gravitational force which drives the adatoms to locations with a lower potential energy. Because the adatoms mass is usually very small we can neglect this contribution and assume $\mu = 0$.

The higher order diffusion which is proportional to ν is, as we have said, a description of the random adatoms dispersal which try to minimize the system chemical potential by increasing the number of bonds among adatoms and the crystal. Finally, the nonlinearity, which is proportional to λ, is the Monge-Ampère differential operator. This nonlinear term might be associated to a Monge-Kantorovich optimal rearrangement of the adatoms on the surface.

I would like to mention other models of fourth order trying to describe epitaxial growth, for instance the variational model considered by Kohn and Yan [94] and Kohn and Otto [93] and the nonvariational model considered by Winkler [147]. In the references of these papers the reader can find the theoretical physics modelization. There exist another kind of models taking into account the bulk of the consolidated crystal under the interface, see [73, 74] and the references therein.

4.1.2 The Mathematical Problem

We have found the following initial-boundary value problem:

$$\begin{cases} u_t + \Delta^2 u = \det(D^2 u) + \lambda f & x \in \Omega\,,\ t > 0\,, \\ u(x, 0) = u_0(x), & x \in \Omega\,, \\ \text{boundary conditions} & x \in \partial\Omega\,,\ t > 0\,, \end{cases} \tag{4.5}$$

where $\lambda \in \mathbb{R}$, and f is some function possibly depending on both space and time coordinates and belonging to some Lebesgue space. The initial condition $u_0(x)$ is also assumed to belong to some Sobolev space. We will consider the following sets of boundary conditions

$$u = u_\nu = 0, \qquad x \in \partial\Omega,$$

which we will refer to as Dirichlet boundary conditions, and

$$u = \Delta u = 0, \qquad x \in \partial\Omega,$$

which we will refer to as Navier boundary conditions.

We will start with the associated stationary problem

$$\begin{cases} \Delta^2 u = \det\left(D^2 u\right) + \lambda f, & x \in \Omega \subset \mathbb{R}^2, \\ \text{boundary} \quad \text{conditions}, \end{cases} \tag{4.6}$$

where Ω has smooth boundary, n is the unit outward normal to $\partial\Omega, f$ is a function with a suitable hypothesis of summability and $\lambda > 0$. We will concentrate on Dirichlet boundary conditions and Navier conditions.

4.1.3 Functional Setting

We begin by studying some properties of the nonlinear term. If v is a smooth function we have in a elementary way the following chain of equalities,

$$\det\left(D^2 v\right) = v_{x_1x_1} v_{x_2x_2} - v_{x_1x_2}^2 = (v_{x_1} v_{x_2x_2})_{x_1} - (v_{x_1} v_{x_2x_1})_{x_2} =$$
$$(v_{x_1} v_{x_2})_{x_1x_2} - \tfrac{1}{2}(v_{x_2}^2)_{x_1x_1} - \tfrac{1}{2}(v_{x_1}^2)_{x_2x_2}.$$

From now on we will assume $\Omega \subset \mathbb{R}^2$ is open, bounded and has a smooth boundary.

Notice that if $v \in W^{2,2}(\Omega)$ by density we can consider the above identities in $\mathcal{D}'(\Omega)$, the space of distributions. This subject is deeply related with a conjecture by J. Ball:

If $u = (u^1, u^2) \in W^{1,p}(\Omega, \mathbb{R}^2)$ *consider* $\det(Du) = u_{x_1}^1 u_{x_2}^2 - u_{x_2}^1 u_{x_1}^2$ *and define*

$$\mathrm{Det}(Du) = (u^1 u_{x_2}^2)_{x_1} - (u^1 u_{x_1}^2)_{x_2}.$$

When is it true that $\det(Du) = \mathrm{Det}(Du)$? A positive answer, among others results, was given by S. Müller in [111] who proves that the answer is affirmative if $p \geq \dfrac{4}{3}$.

We will use Theorem VII.2, page 278 by Coifman, Lions, Meyer and Semmes in [53], that we formulate as follows.

Lemma 4.1 *Let* $v \in W^{2,2}(\mathbb{R}^2)$. *Then,*

$$det\left(D^2 v\right),$$

$$(v_{x_1} v_{x_2x_2})_{x_1} - (v_{x_1} v_{x_2x_1})_{x_2}$$

and

$$(v_{x_1}v_{x_2})_{x_1x_2} - \frac{1}{2}(v_{x_2}^2)_{x_1x_1} - \frac{1}{2}(v_{x_1}^2)_{x_2x_2}$$

belong to the space $\mathcal{H}^1(\mathbb{R}^2)$ *and are equal in it, where* $\mathcal{H}^1(\mathbb{R}^2)$ *is the Hardy space.*

All the expressions involving third derivatives are understood in the distributional sense; then the result in the lemma is highly non-trivial and for the proof we refer to [53]. It is interesting to point out that this result deeply depends on Luc Tartar and François Murat arguments on *compactness by compensation*, see [141, 142] and [112] respectively.

For the reader convenience we recall the definition of the Hardy space in $\mathbb{R}^N$ (see Stein and Weiss [140]).

Definition 4.2 The Hardy space in $\mathbb{R}^N$ is defined in an equivalent way as follows

$$\begin{aligned}\mathcal{H}^1(\mathbb{R}^N) = \{f \in L^1(\mathbb{R}^N) \mid R_j(f) \in L^1(\mathbb{R}^N),\ j = 1, 2, \cdots, N\} = \\ \{f \in L^1(\mathbb{R}^N) \mid \sup |f * h_t(x)| \in L^1(\mathbb{R}^N)\},\end{aligned}$$

where R_j is the classical Riesz transform, that is,

$$R_j(f)(x) = V.P.(K * f)(x), \quad K(y) = C_n \frac{y_j}{|y|^{N+1}}, \ j = 1, 2, \cdots, N,$$

and

$$h_t(x) = \frac{1}{t^N} h\left(\frac{x}{t}\right), \text{ where } h \in C_0^\infty(\mathbb{R}^N),\ h(x) \geq 0 \text{ and } \int_{\mathbb{R}^N} h\,dx = 1.$$

Remark 4.3 The following two properties hold:

1. $\sum_{k=1}^N R_k^2 = Id.$
2. $R_j = \dfrac{\partial}{\partial x_j}(-\Delta)^{-\frac{1}{2}}.$

See [47] and [48].

Notice that, as a direct consequence of the definition, if $f \in \mathcal{H}^1(\mathbb{R}^N)$ and $E_{1,N}(x)$ is the fundamental solution to the Laplacian in $\mathbb{R}^N$, then

$$u(x) = \int_{\mathbb{R}^N} E_{1,N}(x - y) f(y) dy$$

verifies that $u \in W^{2,1}(\mathbb{R}^N)$. See for instance [139].

In a similar way if we consider $E_{2,N}(x)$, the fundamental solution to Δ^2 in $\mathbb{R}^N$, a direct calculation shows that $D^\alpha E(x)$, $|\alpha| = 4$, are of the form

$$D^\alpha E_{2,N}(x) = \frac{H_\alpha(\bar{x})}{|x|^N}, \quad \bar{x} = \frac{x}{|x|}$$

where H_α is a positively homogeneous function of zero degree and

$$\int_{S^{N-1}} H_\alpha(\bar{x}) d\bar{x} = 0,$$

that is, $D^\alpha E_{2,N}(x)$ is a classical Calderon-Zygmund kernel. For $f \in \mathcal{H}^1(\mathbb{R}^N)$ and zero outside of a bounded set consider

$$u(x) = \int_{\mathbb{R}^N} E_{2,N}(x-y) f(y) dy,$$

then $u \in W^{4,1}(\mathbb{R}^N)$. D.C. Chang, G. Dafni, and E.M. Stein in [51] give the definition of Hardy space in a bounded domain Ω in order to have the regularity theory for the Laplacian similar to the one in $\mathbb{R}^N$.

The extension of this kind of regularity result to the bi-harmonic equation on bounded domains is of interest for the current problem.

We use the following consequence which is a by product of the results in [53] and in [51].

Lemma 4.4 *Let $u \in W_0^{2,2}(\Omega)$. Then*

$$Det\left(D^2 u\right) = (u_{x_1} u_{x_2})_{x_1 x_2} - \frac{1}{2}(u_{x_2}^2)_{x_1 x_1} - \frac{1}{2}(u_{x_1}^2)_{x_2 x_2}$$

in $L^1(\Omega) \cap h_r^1(\Omega)$. Here $h_r^1(\Omega)$ is the class of function restrictions of $\mathcal{H}^1(\mathbb{R}^N)$ to Ω.

4.1.4 Lagrangian for Det $(D^2 u)$: The Dirichlet Conditions

We will try to obtain a Lagrangian for which the Euler first variation is the determinant of the Hessian matrix. Our ingredients will be the distributional identity

$$\det\left(D^2 v\right) = (v_{x_1} v_{x_2})_{x_1 x_2} - \frac{1}{2}(v_{x_2}^2)_{x_1 x_1} - \frac{1}{2}(v_{x_1}^2)_{x_2 x_2}$$

and the fact that $C_0^\infty(\Omega)$ is dense in $W_0^{2,2}(\Omega)$.

Consider $\phi \in \mathcal{C}_0^\infty(\Omega)$ and

$$\int_\Omega \det\left(D^2u\right)\phi\,dx = \int_\Omega \left[-\frac{1}{2}(u_{x_2}^2)_{x_1x_1} - \frac{1}{2}(u_{x_1}^2)_{x_2x_2} + (u_{x_1}u_{x_2})_{x_1x_2}\right]\phi\,dx$$

$$= \int_\Omega \left[\frac{1}{2}\phi_{x_1}(u_{x_2}^2)_{x_1} + \frac{1}{2}\phi_{x_2}(u_{x_1}^2)_{x_2} + u_{x_1}u_{x_2}\phi_{x_1x_2}\right]dx$$

$$= \frac{d}{dt}G(u+t\phi)\bigg|_{t=0},$$

where

$$G(u) := \int_\Omega u_{x_1}u_{x_2}u_{x_1x_2}\,dx. \tag{4.7}$$

Notice that by density we can take $\phi \in W_0^{2,2}(\Omega)$ and by direct application of Lemma 4.1 above we find that the first variation of $G(u)$ on $W_0^{2,2}(\Omega)$ is

$$\frac{\delta G(u)}{\delta u} = \det\left(D^2u\right).$$

As a consequence we can prove that

$$\int_\Omega u\det(D^2u) = 3\int_\Omega u_xu_yu_{xy} \quad \forall u \in W_0^{2,2}(\Omega). \tag{4.8}$$

Then we will consider as *energy functional* for problem (4.6) with Dirichlet the following one

$$J_\lambda(u) = \frac{1}{2}\int_\Omega |\Delta u|^2\,dx - \int_\Omega u_{x_1}u_{x_2}u_{x_1x_2}\,dx - \lambda\int_\Omega fu\,dx, \tag{4.9}$$

defined in $W_0^{2,2}(\Omega)$.

Remark 4.5 Notice that this Lagrangian is not useful for other boundary conditions. Indeed, consider $\phi \in \mathcal{X} = \{\phi \in \mathcal{C}^\infty(\Omega) \mid \phi(x) = 0 \text{ on } \partial\Omega\}$ and u a smooth function, then

$$\frac{d}{dt}G(u+t\phi)|_{t=0} = \int_\Omega \left(u_{x_1}u_{x_2}\phi_{x_1x_2} + u_{x_1}\phi_{x_2}u_{x_1x_2} + \phi_{x_1}u_{x_2}u_{x_1x_2}\right)dx$$

$$= \int_\Omega \det\left(D^2u\right)\phi\,dx - \frac{1}{2}\int_{\partial\Omega} u_{x_1}u_{x_2}\left(\phi_{x_1}\nu_{x_2} + \phi_{x_2}\nu_{x_1}\right)ds,$$

and therefore for $\phi \in \mathcal{X}$ the boundary term does not cancel.

This observation justifies the dependence of the problem on the boundary conditions.

4.2 The Stationary Problem with Dirichlet Conditions

We will study the following problem

$$\begin{cases} \Delta^2 u = \det\left(D^2 u\right) + \lambda f, & x \in \Omega \subset \mathbb{R}^2, \\ \text{u=0}, \quad \dfrac{\partial u}{\partial n} = 0 \text{ on } \partial\Omega, \end{cases} \tag{4.10}$$

this is, Dirichlet boundary conditions, where Ω is a bounded domain with smooth boundary and $f \in L^1(\Omega)$. The natural framework is the space $W_0^{2,2}(\Omega)$, that is the completion of $C_0^\infty(\Omega)$ with the norm of $W^{2,2}(\Omega)$. According with the previous paragraph, in order to find solution to problem (4.10) we will find critical points to the functional of energy

$$J_\lambda(u) = \frac{1}{2}\int_\Omega |\Delta u|^2\, dx - \int_\Omega u_{x_1} u_{x_2} u_{x_1 x_2}\, dx - \lambda \int_\Omega fu\, dx, \tag{4.11}$$

that clearly is well defined in $W_0^{2,2}(\Omega)$.

Remark 4.6 A quite similar variational structure is verified for periodic conditions. We skip the details about the periodic problem.

4.2.1 Variational Approach to the Problem with Dirichlet Condition

Since J_λ is unbounded from below we can not minimize. Then we will try to find solutions as critical points of the functional J_λ, precisely as a consequence of the geometry of J_λ, of *mountain-pass* type.

The Geometry of J_λ Notice that by Hölder and Sobolev inequalities we find the following estimate

$$\begin{aligned} &J_\lambda(u) \\ &\geq \\ &\frac{1}{2}\int_\Omega |\Delta u|^2\, dx - \left(\int_\Omega |u_{x_1 x_2}|^2\, dx\right)^{\frac{1}{2}} \left(\int_\Omega |u_{x_1}|^4\, dx\right)^{\frac{1}{4}} \left(\int_\Omega |u_{x_2}|^4\, dx\right)^{\frac{1}{4}} \\ &- \lambda ||f||_1 ||u||_\infty \\ &\geq \end{aligned}$$

$$\frac{1}{2}\int_\Omega |\Delta u|^2\,dx - c_1\left(\int_\Omega |\Delta u|^2\,dx\right)^{\frac{3}{2}} - \lambda c_2\|f\|_1\left(\int_\Omega |\Delta u|^2\,dx\right)^{\frac{1}{2}}$$

$$\equiv g\left(\|\Delta u\|_2\right),$$

where

$$g(s) = \frac{1}{2}s^2 - c_1 s^3 - \lambda\, c_2\, \|f\|_1\, s. \qquad (4.12)$$

Therefore we easily prove that for $0 < \lambda < \lambda_0$ small enough, the *radial lower estimate (in the Sobolev space)*, given by g has a negative local minimum and a positive local maximum. Moreover, it is easy to check that:

1. There exists a function $\phi \in W_0^{2,2}(\Omega)$ such that

$$\int_\Omega f\phi\,dx > 0.$$

2. There exists a function $\psi \in W_0^{2,2}(\Omega)$ such that

$$\int_\Omega \psi_{x_1}\psi_{x_2}\psi_{x_1x_2}\,dx > 0.$$

For the function ϕ we just need it to be a local mollification of f. The case of ψ is a bit more involved but one still has many possibilities such as

$$\psi = [(1-|x|^2)^+]^4,$$

where $|x| = \sqrt{x_1^2 + x_2^2}$ and $(\cdot)^+ = \max\{\cdot, 0\}$, that fulfils the positivity criterion even pointwise in a domain containing the unit ball. Then, in general, if $B_{2r}(x_0) \subset \Omega$ we consider $\psi_\Omega(x) = \psi(\dfrac{x-x_0}{r})$. Other suitable functions can be found by means of deforming this one adequately. Notice that the ψ function we have chosen is in $C^2(\mathbb{R}^2)$.

According to the previous remark we find that

$$J_\lambda(t\phi) < 0 \text{ for } t \text{ small enough and } J_\lambda(s\psi) < 0 \text{ for } s \text{ large enough.}$$

This behavior and the *radial minorant* in $W_0^{2,2}(\Omega)$, suggests a kind of *mountain pass* geometry. See the classical paper by Ambrosetti and Rabinowitz [15].

Palais-Smale Condition for J_λ As usual, we call $\{u_k\}_{k\in\mathbb{N}} \subset W_0^{2,2}(\Omega)$ a Palais-Smale sequence for J_λ to the level c if

(i) $J_\lambda(u_k) \to c$ as $k \to \infty$
(ii) $J'_\lambda(u_k) \to 0$ in $W^{-2,2}(\Omega)$.

We say that J_λ satisfies the local Palais-Smale condition to the level c if each Palais-Smale sequence to the level c, $\{u_k\}_{k\in\mathbb{N}}$, admits a strongly convergent subsequence in $W_0^{2,2}(\Omega)$.

We are able to prove the following compactness result.

Lemma 4.7 *Assume a bounded Palais-Smale condition for J_λ, that is $\{u_k\}_{k\in\mathbb{N}} \subset W_0^{2,2}(\Omega)$ verifying*

$$1)\quad J_\lambda(u_k) \to c \text{ as } k \to \infty; \qquad 2)\quad J'_\lambda(u_k) \to 0 \text{ in } W^{-2,2}.$$

Then there exists a subsequence $\{u_k\}_{k\in\mathbb{N}}$ that converges in $W_0^{2,2}(\Omega)$.

Proof Since $\{u_k\}_{k\in\mathbb{N}} \subset W_0^{2,2}(\Omega)$ is bounded, up to passing to a subsequence, we have:

(i) $u_k \rightharpoonup u$ weakly in $W_0^{2,2}(\Omega)$,
(ii) $\nabla u_k \to \nabla u$ strongly in $[L^p(\Omega)]^2$ for all $p < \infty$,
(iii) $u_k \to u$ uniformly in Ω.

We can write the condition $J'_\lambda(u_k) \to 0$ in $W^{-2,2}$ as

$$\Delta^2 u_k = \det(D^2 u_k) + \lambda f + y_k, \;\; u_k \in W_0^{2,2}(\Omega) \text{ and } y_k \to 0 \text{ in } W^{-2,2}(\Omega). \tag{4.13}$$

Notice that multiplying (4.13) by $(u_k - u)$, we have for all fixed k

$$\begin{gathered}\int_\Omega \Delta(u_k)\Delta(u_k - u)\,dx = \\ \int_\Omega (u_k - u)\det\left(D^2 u_k\right)\,dx + \lambda\int_\Omega f\,(u_k - u)\,dx + y_k(u_k - u).\end{gathered} \tag{4.14}$$

The three terms on the right hand side go to zero as $k \to \infty$ by the convergence properties *i)* and *iii)*. Moreover adding in both terms of (4.14)

$$-\int_\Omega \Delta u\,\Delta(u_k - u)\,dx = o(1) \quad k \to \infty,$$

we obtain,

$$\int_\Omega |\Delta(u_k - u)|^2\,dx = \int_\Omega (u_k - u)\det\left(D^2 u_k\right)\,dx + \lambda\int_\Omega f\,(u_k - u)\,dx +$$

$$y_k(u_k - u) - \int_\Omega \Delta u\,\Delta(u_k - u)\,dx.$$

As a consequence

$$\int_{\Omega} |\Delta(u_k - u)|^2\, dx \to 0 \text{ as } k \to \infty, \tag{4.15}$$

that is, J_λ satisfies the Palais-Smale condition to the level c. □

4.2.2 The Multiplicity Result

We prove the following existence and multiplicity result.

Theorem 4.8 *Let $\Omega \subset \mathbb{R}^2$ be a bounded domain with smooth boundary. Consider $f \in L^1(\Omega)$ and $\lambda > 0$. Then there exists a λ_0 such that for $0 < \lambda < \lambda_0$ problem* (4.10) *has at least two solutions.*

Proof By the Sobolev embedding theorem the functional J_λ is well defined in $W_0^{2,2}(\Omega)$, is continuous and Gateaux differentiable, and its derivative is weak-* continuous (precisely the regularity required in the weak version by Ekeland of the *mountain pass theorem* in [17]).

We will try to prove the existence of a solution which corresponds to a negative local minimum of J_λ and a solution which corresponds to a positive mountain pass level of J_λ.

Step 1 J_λ has a local minimum u_0, such that $J_\lambda(u_0) < 0$.

We use the ideas in [77] to solve problems with concave-convex semilinear nonlinearities.

Consider $\lambda_0 > 0$ such that, if $0 < \lambda < \lambda_0$ and consider g_λ defined in (4.12) for the values of λ such that g_λ attaints its positive maximum at a $r_{max} > 0$ and for s close to zero $g_\lambda(s) < 0$. Let r_0 be, the lower positive zero of g_λ and $r_0 < r_1 < r_{max} < r_2$ such that $g_\lambda(r_1) > 0$, $g_\lambda(r_2) > 0$. Now consider a cutoff function

$$\tau : \mathbb{R}_+ \to [0, 1],$$

such that τ is nonincreasing, $\tau \in \mathcal{C}^\infty$ and it verifies

$$\begin{cases} \tau(s) = 1 & \text{if} \quad s \le r_0, \\ \tau(s) = 0 & \text{if} \quad s \ge r_1. \end{cases}$$

Let $\Theta(u) = \tau(\|\Delta u\|_2)$. We consider the truncated functional

$$F_\lambda(u) = \frac{1}{2}\int_{\Omega} |\Delta u|^2\, dx - \int_{\Omega} u_{x_1} u_{x_2} u_{x_1 x_2} \Theta(u)\, dx - \lambda \int_{\Omega} fu\, dx. \tag{4.16}$$

As above, by Hölder and Sobolev inequalities we see $F(u) \geq h(\|\Delta u\|_2)$, with

$$h(s) = \frac{1}{2}\, s^2 - c_1\, \tau(s)\, s^3 - \lambda\, ||f||_1\, c_2\, s.$$

Lemma 4.9 *The properties of F defined by* (4.16) *are the following.*

1. *F_λ has the same regularity as J_λ.*
2. *If $F_\lambda(u) < 0$, then $\|\Delta u\|_2 < r_0$, and $F_\lambda(u) = J_\lambda(u)$ if $\|\Delta u\|_2 < r_0$.*
3. *Let m be defined by $m = \inf\limits_{v \in W_0^{2,2}(\Omega)} F_\lambda(v)$.*

Then F_λ verifies a local Palais-Smale condition to the level m.

Proof (1) and (2) are immediate. To prove (3), observe that all Palais-Smale sequences of minimizers of F_λ, since $m < 0$, must be bounded. Then by Lemma 4.7 we conclude. □

Observe that, by (2), if we find some negative critical value for F_λ, then we have that m is a negative critical value of J_λ and there exist u_0 local minimum for J_λ.

Step 2 If λ is small enough, J_λ has a mountain pass *critical point, u_*, such that $J_\lambda(u_*) > 0$.*

By the estimates in Sect. 4.2.1, J_λ verifies the geometrical requirements of the Mountain Pass Theorem (see [15] and [17]). Consider u_0 the local minimum such that $J_\lambda(u_0) < 0$ and consider $v \in W_0^{2,2}(\Omega)$ with $||\Delta v||_2 > r_{max}$ and such that $J_\lambda(v) < J_\lambda(u_0)$. We define

$$\Gamma = \{\gamma \in C\left([0,1], W_0^{2,2}(\Omega)\right) \mid \gamma(0) = u_0,\ \gamma(1) = v\},$$

and the minimax value

$$c = \inf_{\gamma \in \Gamma} \max_{t \in [0,1]} J_\lambda[\gamma(t)].$$

Applying the Ekeland variational principle (see [64]), there exists a Palais-Smale sequence to the level c, i. e. there exists $\{u_k\}_{k \in \mathbb{N}} \subset W_0^{2,2}(\Omega)$ such that

$$(1) \quad J_\lambda(u_k) \to c \text{ as } k \to \infty; \quad 2) \quad J'_\lambda(u_k) \to 0 \text{ in } W^{-2,2}.$$

Claim If $\{u_k\}_{k \in \mathbb{N}} \subset W_0^{2,2}(\Omega)$ is a Palais-Smale sequence for J_λ at the level c, then there exists $C > 0$ such that $||\Delta u_k||_2 < C$.

Since the results in Sect. 4.1.3 hold then if $u \in W_0^{2,2}(\Omega)$, integrating by parts we find that

$$\begin{aligned}\int_\Omega u \det\left(D^2 u\right) dx &= \int_\Omega u\left[(u_{x_1}\, u_{x_2x_2})_{x_1} - (u_{x_1}\, u_{x_2x_1})_{x_2}\right] dx \\ &\quad - \int_\Omega (u_{x_1})^2\, u_{x_2x_2}\, dx + \int_\Omega u_{x_1}\, u_{x_2x_1}\, u_{x_2}\, dx \\ &= 2\int_\Omega u_{x_1}\, u_{x_1x_2}\, u_{x_2}\, dx + \int_\Omega u_{x_1}\, u_{x_2x_1}\, u_{x_2}\, dx \\ &= 3\int_\Omega u_{x_1}\, u_{x_2x_1}\, u_{x_2}\, dx.\end{aligned} \tag{4.17}$$

Then if $\{u_k\}_{k\in\mathbb{N}} \subset W_0^{2,2}(\Omega)$ is a Palais-Smale sequence for J_λ at the level c and calling $\langle y_k, u_k\rangle = \langle J'_\lambda(u_k), u_k\rangle$

$$\begin{aligned}c + o(1) &= J_\lambda(u_k) - \frac{1}{3}\langle J'_\lambda(u_k), u_k\rangle + \frac{1}{3}\langle y_k, u_k\rangle \\ &\geq \left(\frac{1}{2} - \frac{1}{3}\right)\int_\Omega |\Delta u_k|^2\, dx - \frac{1}{3}\,||y_k||_{H^{-2}}\left(\int_\Omega |\Delta u_k|^2\, dx\right)^{\frac{1}{2}} \\ &-\frac{2}{3}\,\lambda\, C_S\, ||f||_{L^1}\left(\int_\Omega |\Delta u_k|^2\, dx\right)^{\frac{1}{2}},\end{aligned}$$

where C_S is a suitable Sobolev constant. This inequality implies that the sequence is bounded.

By using Lemma 4.7, J_λ satisfies the Palais-Smale condition to the level c. Therefore

1. $J_\lambda(u_*) = \lim_{k\to\infty} J_\lambda(u_k) = c$ (and then u_* is different from the local minimum, as in this case the value of the functional at this point is positive while in the other one was negative).
2. $J'_\lambda(u_*) = 0$, thus

$$\Delta^2 u_* = \det(D^2 u_*) + \lambda f,\ \ u_* \in W_0^{2,2}(\Omega).$$

In other words u_* is a *mountain pass type* solution to the problem (4.10). □

Remark 4.10 If $\lambda = 0$ the local minimum is the trivial solution, nevertheless the mountain-pass solution remains as a nontrivial solution to the problem.

An analytical observation is that we cannot directly conclude that a bounded Palais-Smale sequence gives a solution in the distributional sense; indeed, we would need the convergence property

$$\det(D^2 u_k) \rightharpoonup \det(D^2 u_*) \text{ at least in } L^1(\Omega).$$

To have this property up to passing to a subsequence we need almost everywhere convergence (see the result by Jones and Journé in [88]). Notice that a. e. convergence for the second derivatives is only known after the proof of Lemma 4.7.

4.2.3 Regularity and Some Extensions

Consider now the nonhomogeneous problem

$$\begin{cases} \Delta^2 u = \det(D^2 u) + f & \text{in } \Omega \\ u = g & \text{on } \partial\Omega \\ u_\nu = h & \text{on } \partial\Omega \end{cases}, \tag{4.18}$$

where $f \in L^1(\Omega)$, $g \in W^{3/2,2}(\partial\Omega)$, $h \in H^{1/2,2}(\partial\Omega)$. The following result holds.

Theorem 4.11 *There exists $\gamma > 0$ such that if*

$$\|f\|_1 + \|g\|_{W^{3/2,2}(\partial\Omega)} + \|h\|_{W^{1/2,2}(\partial\Omega)} < \gamma \tag{4.19}$$

then (4.18) *admits at least two weak solutions in $W^{2,2}(\Omega)$, a stable solution and a mountain pass solution.*

Proof Since the embedding $L^1(\Omega) \subset W^{-2,2}(\Omega)$ holds (see for instance [78, Theorem 2.16]) the linear problem

$$\begin{cases} \Delta^2 v = f & \text{in } \Omega \\ v = g & \text{on } \partial\Omega \\ v_\nu = h & \text{on } \partial\Omega. \end{cases} \tag{4.20}$$

admits a unique weak solution $v \in W^{2,2}(\Omega)$ which moreover satisfies

$$\|D^2 v\|_2 \le C\Big(\|f\|_1 + \|g\|_{W^{3/2,2}(\partial\Omega)} + \|h\|_{W^{1/2,2}(\partial\Omega)}\Big) \tag{4.21}$$

for some $C > 0$ independent of f, g, h. Define $w = u - v$, then w solves the problem

$$\begin{cases} \Delta^2 w = \det[D^2(w + v)] & \text{in } \Omega \\ w = w_\nu = 0 & \text{on } \partial\Omega. \end{cases}$$

This problem can be written as

$$\begin{cases} \Delta^2 w = \det(D^2 w) + \det(D^2 v) + v_{xx} w_{yy} + w_{xx} v_{yy} - 2 w_{xy} v_{xy} & \text{in } \Omega \\ w = w_\nu = 0 & \text{on } \partial\Omega. \end{cases} \tag{4.22}$$

For $\|D^2 v\|_2$ small enough we find a mountain pass geometry for the associated energy functional. In view of (4.21), the mountain pass geometry is ensured if γ in (4.19) is sufficiently small. This geometry yields the existence of a locally minimum solution and of a mountain pass solution. □

By classical regularity results in [10] and [11] (see too [78]), we have the following statement.

Theorem 4.12 *Assume that, for some integer $k \geq 0$ we have: $\partial\Omega \in C^{k+4}$, $f \in W^{k,2}(\Omega)$, $g \in W^{k+7/2,2}(\partial\Omega)$, $h \in W^{k+5/2,2}(\partial\Omega)$. Then any solution to* (4.18) *satisfies*

$$u \in W^{k+4,2}(\Omega) .$$

In particular, any solution to

$$\begin{cases} \Delta^2 u = \det(D^2 u) & \text{in } \Omega \\ u = u_\nu = 0 & \text{on } \partial\Omega \end{cases}$$

is as smooth as the boundary permits.

See [67] for some details.

Remark 4.13 Since we are working in dimension $N = 2$, if we stop the previous proof at the first step, we see that, in a C^3 domain, any solution to

$$\begin{cases} \Delta^2 u = \det(D^2 u) + f & \text{in } \Omega \\ u = u_\nu = 0 & \text{on } \partial\Omega \end{cases}$$

with $f \in L^1(\Omega)$ belongs to $W^{r,2}(\Omega)$ for any $r < 3$.

4.2.4 The Nehari Manifold and the Mountain Pass Level

Some geometrical aspect of the *mountain-pass level* will be useful to understand the asymptotic behavior of the parabolic problems in some cases. The energy functional for the stationary problem

$$\begin{cases} \Delta^2 u = \det(D^2 u) & \text{in } \Omega \\ u = u_\nu = 0 & \text{on } \partial\Omega \end{cases} \tag{4.23}$$

is

$$J(v)=\frac{1}{2}\int_{\Omega}|\Delta v|^2-\int_{\Omega}v_xv_yv_{xy}\qquad \forall v\in W_0^{2,2}(\Omega). \tag{4.24}$$

It is shown in Sect. 4.3 that J has a mountain pass geometry and that the corresponding mountain pass level is given by

$$d=\inf_{\gamma\in\Gamma}\max_{0\le s\le 1}J(\gamma(s)) \tag{4.25}$$

where $\Gamma:=\{\gamma\in C([0,1],W_0^{2,2}(\Omega));\ \gamma(0)=0,\ J(\gamma(1))<0\}$. We aim to characterize differently d and to relate it with the so-called Nehari manifold defined by

$$\mathcal{N}:=\left\{v\in W_0^{2,2}(\Omega)\setminus\{0\};\ \langle J'(v),v\rangle=\|v\|^2-3\int_{\Omega}v_xv_yv_{xy}=0\right\}$$

where $\langle\cdot,\cdot\rangle$ denotes the duality pairing between $W^{2,-2}(\Omega)$ and $W_0^{2,2}(\Omega)$. To this end, we introduce the set

$$B:=\{v\in W_0^{2,2}(\Omega);\ \int_{\Omega}v_xv_yv_{xy}=1\}. \tag{4.26}$$

It is clear that $v\in\mathcal{N}$ if and only if $\alpha v\in B$ for some $\alpha>0$. In particular, not on all the straight directions starting from 0 in the phase space $W_0^{2,2}(\Omega)$ there exists an intersection with $\mathcal{N}$. Hence, $\mathcal{N}$ is an unbounded manifold (of codimension 1) which separates the two regions

$$\mathcal{N}_+=\left\{v\in W_0^{2,2}(\Omega);\ \|v\|^2>3\int_{\Omega}v_xv_yv_{xy}\right\}$$

and

$$\mathcal{N}_-=\left\{v\in W_0^{2,2}(\Omega);\ \|v\|^2<3\int_{\Omega}v_xv_yv_{xy}\right\}.$$

A further functional needed in the sequel is given by

$$I(v)=\int_{\Omega}v_xv_yv_{xy}\,. \tag{4.27}$$

In Fig. 1 we sketch a geometric representation of the Nehari manifold $\mathcal{N}$ which summarizes the results obtained in the present section.

The next result states some properties of $\mathcal{N}_\pm$.

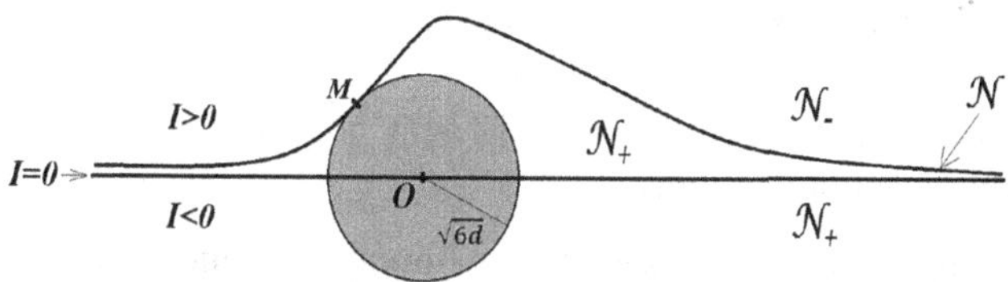

Fig. 1 The phase space $W_0^{2,2}(\Omega)$ with: $\mathcal{N}$ = Nehari manifold, M = mountain pass point, and I given by (4.27)

Theorem 4.14 *Let $v \in W_0^{2,2}(\Omega)$, then the following implications hold:*

(i) $0 < \|v\|^2 < 6d \implies v \in \mathcal{N}_+$;
(ii) $v \in \mathcal{N}_+,\ J(v) < d \implies 0 < \|v\|^2 < 6d$;
(iii) $v \in \mathcal{N}_- \implies \|v\|^2 > 6d$.

Proof It is well-known [15] that the mountain pass level d may also be defined by

$$d = \min_{v \in \mathcal{N}} J(v) . \tag{4.28}$$

Using (4.28) and the definition of $\mathcal{N}$ we obtain

$$d = \min_{v \in \mathcal{N}} J(v) = \min_{v \in \mathcal{N}} \left(\frac{\|v\|^2}{2} - \int_\Omega v_x v_y v_{xy} \right) = \min_{v \in \mathcal{N}} \frac{\|v\|^2}{6}$$

which proves (i) since $\mathcal{N}$ separates $\mathcal{N}_+$ and $\mathcal{N}_-$.

If $v \in \mathcal{N}_+$, then $-\int_\Omega v_x v_y v_{xy} > -\|v\|^2/3$. If $J(v) < d$, then $\|v\|^2 - 2\int_\Omega v_x v_y v_{xy} < 2d$. By combining these two inequalities we obtain (ii).

Finally, recalling the definitions of $\mathcal{N}_\pm$, (iii) follows directly from (i). □

We provide a different characterization of the mountain pass level. For the proof see [67].

Theorem 4.15 *The mountain pass level d for J is also determined by*

$$d = \min_{v \in B} \frac{\|v\|^6}{54} . \tag{4.29}$$

Moreover, d can be estimated from below by

$$d \geq \frac{8}{27} \min_{W_0^{2,2}(\Omega)} \frac{\left(\int_\Omega |\Delta v|^2\right)^2}{\int_\Omega |\nabla v|^4} = \frac{8}{27} S ,$$

(that is, S is the best constant for the compact embedding $W_0^{2,2}(\Omega) \subset W_0^{1,4}(\Omega)$).

4.2.5 Further Results for the Elliptic Problems

We will state some results involving different boundary condition and related inequalities in higher dimensions.

Some Existence Results Including Navier Boundary Conditions We will find a solution to our problem with Navier boundary conditions

$$\begin{cases} \qquad \Delta^2 u = \det\left(D^2 u\right) + \lambda f, \; x \in \Omega \subset \mathbb{R}^2, \\ u = 0, \quad \Delta u = 0 \qquad\qquad\qquad \text{on } \partial\Omega, \end{cases} \tag{4.30}$$

and also with Dirichlet boundary conditions

$$\begin{cases} \qquad \Delta^2 u = \det\left(D^2 u\right) + \lambda f, \qquad x \in \Omega \subset \mathbb{R}^2, \\ u = 0, \quad \dfrac{\partial u}{\partial n} = 0 \text{ on } \partial\Omega. \end{cases} \tag{4.31}$$

We will prove the existence of at least one solution to problems (4.30) and (4.31) for a datum small enough, by means of fixed point methods. First of all we need the following technical result that involves the structure of the nonlinearity and some cancelation properties.

Lemma 4.16 *For any functions $v_1, v_2 \in W^{1,2}(\Omega)$ and $v_3 \in W_0^{1,2}(\Omega) \cap W^{2,2}(\Omega)$ the following equality is fulfilled*

$$\int det\,(\nabla v_1, \nabla v_2)\, v_3\, dx = \int v_1\, \nabla v_2 \cdot \nabla^{\perp} v_3\, dx, \tag{4.32}$$

where $\nabla^{\perp} v_3 = \left(\partial_{x_2} v_3, -\partial_{x_1} v_3\right)$.

The main result in this setting is the next.

Theorem 4.17 *If $\lambda > 0$ is small enough then:*

(a) There exists $u \in W_0^{1,2}(\Omega) \cap W^{2,2}(\Omega)$ solution to problem (4.30).
(b) There exists $u \in W_0^{2,2}(\Omega)$ solution to problem (4.31).

The proof and other related questions can be seen in [66].

Problems Involving a 2-Hessian Nonlinearity in Dimension $N = 3$ In [72] the following problem has been studied. Suppose that $\Omega \subset \mathbb{R}^3$ is a bounded smooth set.

Consider the following Dirichlet problem

$$\begin{cases} \Delta^2 u = S_2(D^2 u), \; \Omega, \\ \quad u = 0, \qquad\quad \partial\Omega, \\ \dfrac{\partial u}{\partial n} = 0, \qquad\quad \partial\Omega, \end{cases} \tag{4.33}$$

where S_2 is the 2-nd Hessian operator. Namely we set

$$S_2(D^2u)(x) = \sum_{1\le i<j\le N} \lambda_i(x)\lambda_j(x),$$

where $\lambda_i(x)$ $i = 1, \ldots, N$ is the i-th eigenvalue of the symmetric matrix $D^2u(x)$. We remind that S_2 can also be written in the following way:

$$S_2(D^2u(x)) = \sum_{1\le i<j\le N} \det(D^2_{ij}u(x)),$$

where

$$\det(D^2_{ij}u(x)) = \partial_{ii}u(x)\partial_{jj}u(x) - (\partial_{ij}u(x))^2.$$

This last expression of S_2 allows to formulate the problem as in the case of the Hessian nonlinearity in dimension $N = 2$. The following result is obtained in [72].

Theorem 4.18 *Problem* (4.33) *has at least one nontrivial solution.*

The problem with Navier boundary conditions,

$$\begin{cases} \Delta^2 u = S_2(D^2u) + \lambda u, & \Omega, \\ u = 0, & \partial\Omega, \\ \Delta u = 0, & \partial\Omega. \end{cases} \tag{4.34}$$

is solved by using the bifurcation theory. See [133] and the details for this concrete case in [72].

Theorem 4.19 *Let λ_1 be the first eigenvalue of Δ^2 in Ω with Navier boundary conditions, which is simple. Then, there exists an unbounded continuum of pairs (λ, u), $u \in W^{2,2}(\Omega) \cap W_0^{1,2}(\Omega)$ branching-off from $(\lambda_1, 0)$, where every u is a solution to* (4.34) *with the corresponding λ.*

Moreover, as in the two-dimensional case, by means of Fixed Point methods, we obtain the solvability of the problem with Navier conditions and a source term, that is,

$$\begin{cases} \Delta^2 u = S_2(D^2u) + \mu f(x) & x \in \Omega, \\ u = 0 & x \in \partial\Omega, \\ \Delta u = 0 & x \in \partial\Omega, \end{cases} \tag{4.35}$$

where $\mu > 0$. Precisely, among others, the following result is obtained.

Theorem 4.20 *Let $f \in L^1(\Omega)$. Then, there exists $\mu_0 > 0$ such that for every μ satisfying $0 < \mu < \mu_0$, there exists $u \in W_0^{1,2}(\Omega) \cap W^{2,2}(\Omega)$ solution to problem* (4.35).

4.3 The Parabolic Problem

This section is devoted to the study of the evolution problem

$$u_t + \Delta^2 u = \det(D^2 u) + \lambda f \qquad \text{in } \Omega \times (0, T), \tag{4.36}$$

where $\Omega \subset \mathbb{R}^2$ is a smooth bounded domain and for some $T > 0$. We consider both the sets of boundary conditions $u|_{\partial\Omega} = u_\nu|_{\partial\Omega} = 0$ (Dirichlet condition) and $u|_{\partial\Omega} = \Delta u|_{\partial\Omega} = 0$ (Navier conditions). Here and in the sequel we will be always considering weak solutions.

By using the Galerkin method the following result holds for the associated linear problem.

Theorem 4.21 *Let $0 < T \leq \infty$ and let $f \in L^2(0, T; L^2(\Omega))$. The Dirichlet problem for the linear fourth order parabolic equation*

$$u_t + \Delta^2 u = f \qquad \text{in } \Omega \times (0, T), \tag{4.37}$$

with initial datum $u_0 \in W_0^{2,2}(\Omega)$ admits a unique weak solution in the space

$$C(0, T; W_0^{2,2}(\Omega)) \cap L^2(0, T; W^{4,2}(\Omega)) \cap W^{1,2}(0, T; L^2(\Omega)).$$

The corresponding Navier problem with initial datum $u_0 \in W^{2,2}(\Omega) \cap W_0^{1,2}(\Omega)$ admits a unique weak solution in the space

$$C(0, T; W^{2,2}(\Omega) \cap W_0^{1,2}(\Omega)) \cap L^2(0, T; W^{4,2}(\Omega)) \cap W^{1,2}(0, T; L^2(\Omega)).$$

Furthermore, both cases admit the estimate

$$\max_{0 \leq t \leq T} \|\Delta u\|_2^2 + \int_0^T \|\Delta^2 u\|_2^2 + \int_0^T \|u_t\|_2^2 \leq C\left(\|\Delta u_0\|_2^2 + \int_0^T \|f\|_2^2 \right).$$

The details of the proof can be found in [67].

A fixed point argument and the result above for the linear problem allow us to prove the main result of this section.

Theorem 4.22 *The problem*

$$\begin{cases} u_t + \Delta^2 u = \det(D^2 u) + \lambda f & \text{in } \Omega \times (0, T) \\ u(x, 0) = u_0(x) & \text{in } \Omega \\ u(x, t) = u_\nu(x, t) = 0 & \text{on } \partial\Omega \times (0, T) \end{cases} \tag{4.38}$$

admits a unique solution in

$$\mathcal{X}_T := C(0, T; W_0^{2,2}(\Omega)) \cap L^2(0, T; W^{4,2}(\Omega)) \cap W^{1,2}(0, T; L^2(\Omega)),$$

provided one of the following set of conditions holds

(i) $u_0 \in W_0^{2,2}(\Omega)$, $f \in L^2(0,T;L^2(\Omega))$, $\lambda \in \mathbb{R}$, *and* $T > 0$ *is sufficiently small;*
(ii) $T \in (0,\infty]$, $f \in L^2(0,T;L^2(\Omega))$, *and* $\|u_0\|$ *and* $|\lambda|$ *are sufficiently small.*

Moreover, if $[0,T^*)$ *denotes the maximal interval of continuation of* u *and if* $T^* < \infty$ *then* $\|u(t)\| \to \infty$ *as* $t \to T^*$.

An identical result holds for the Navier problem but this time the solution belongs to the space

$$\mathcal{Y}_T := C(0,T;W^{2,2}(\Omega)\cap W_0^{1,2}(\Omega)) \cap L^2(0,T;W^{4,2}(\Omega)) \cap W^{1,2}(0,T;L^2(\Omega)),$$

assuming that the initial condition $u_0 \in W^{2,2}(\Omega) \cap W_0^{1,2}(\Omega)$.

Proof For all $u \in W^{4,2}(\Omega)$ we have

$$\begin{aligned}\|\det(D^2u)\|_2^2 &= \int_\Omega |\det(D^2u)|^2 \le C\int_\Omega |D^2u|^4 \le C\|D^2u\|_\infty^2 \int_\Omega |D^2u|^2 \\ &\le C\|\Delta u\|_\infty^2\, \|\Delta u\|_2^2 \le C\|\Delta^2 u\|_2^2\, \|\Delta u\|_2^2.\end{aligned}$$

Hence, if $u \in C(0,T;W^{2,2}(\Omega)) \cap L^2(0,T;W^{4,2}(\Omega))$, we may directly estimate

$$\begin{aligned}\|\det(D^2u)\|_{L^2(0,T;L^2(\Omega))}^2 &= \int_0^T \|\det(D^2u)\|_2^2 \le C\int_0^T \|\Delta^2u\|_2^2\, \|\Delta u\|_2^2 \\ &\le C\max_{0\le t\le T} \|\Delta u\|_2^2 \int_0^T \|\Delta^2 u\|_2^2 < \infty\end{aligned}$$

which proves that if $u \in C(0,T;W^{2,2}(\Omega)) \cap L^2(0,T;W^{4,2}(\Omega))$ then $\det(D^2u) \in L^2(0,T;L^2(\Omega))$.

In what follows we focus on the Dirichlet case since the proof for the Navier one follows similarly. We introduce the initial-Dirichlet linear problems

$$\begin{cases}(u_1)_t + \Delta^2 u_1 = \det(D^2v_1) + \lambda f\,, & u_1(x,0) = u_0(x)\,,\\ (u_2)_t + \Delta^2 u_2 = \det(D^2v_2) + \lambda f\,, & u_2(x,0) = u_0(x)\,,\end{cases} \tag{4.39}$$

where $v_1, v_2 \in \mathcal{X}_T$. The just proved inclusion and Theorem 4.21 show that $u_1, u_2 \in \mathcal{X}_T$. Subtracting the equations in (4.39) we get

$$(u_1-u_2)_t + \Delta^2(u_1-u_2) = \det(D^2v_1) - \det(D^2v_2)\,, \quad (u_1-u_2)(x,0) = 0\,,$$

and upon multiplying by $\Delta^2(u_1-u_2)$ and integrating we find

$$\begin{aligned}&(\Delta^2(u_1-u_2), (u_1-u_2)_t)_2 + (\Delta^2(u_1-u_2), \Delta^2(u_1-u_2))_2 \\ &= (\Delta^2(u_1-u_2), \det(D^2v_1) - \det(D^2v_2))_2.\end{aligned}$$

This leads to the inequalities

$$\frac{1}{2}\frac{d}{dt}\|\Delta(u_1-u_2)\|_2^2+\|\Delta^2(u_1-u_2)\|_2^2$$
$$\le \frac{1}{2}\|\Delta^2(u_1-u_2)\|_2^2+\frac{1}{2}\|\det(D^2v_1)-\det(D^2v_2)\|_2^2$$

and, in turn,

$$\frac{d}{dt}\|\Delta(u_1-u_2)\|_2^2+\|\Delta^2(u_1-u_2)\|_2^2\le\|\det(D^2v_1)-\det(D^2v_2)\|_2^2. \tag{4.40}$$

We split the remaining part of the proof into three steps.

Step 1. Existence for Arbitrary Time and Small Data The nonlinear terms can be estimate as follows

$$\|\det(D^2v_1)-\det(D^2v_2)\|_2^2\le C\int_\Omega|D^2(v_1-v_2)|^2(|D^2v_1|+|D^2v_2|)^2\le \tag{4.41}$$

$$C(\|\Delta v_1\|_\infty^2+\|\Delta v_2\|_\infty^2)\|\Delta(v_1-v_2)\|_2^2\le C(\|\Delta^2v_1\|_2^2+\|\Delta^2v_2\|_2^2)\|\Delta(v_1-v_2)\|_2^2,$$

therefore by (4.40) we get

$$\frac{d}{dt}\|\Delta(u_1-u_2)\|_2^2+\|\Delta^2(u_1-u_2)\|_2^2\le C(\|\Delta^2v_1\|_2^2+\|\Delta^2v_2\|_2^2)\|\Delta(v_1-v_2)\|_2^2.$$

Integrating with respect to time we obtain

$$\max_{0\le t\le T}\|\Delta(u_1-u_2)\|_2^2+\int_0^T\|\Delta^2(u_1-u_2)\|_2^2\le \tag{4.42}$$

$$C\max_{0\le t\le T}\|\Delta(v_1-v_2)\|_2^2\int_0^T(\|\Delta^2v_1\|_2^2+\|\Delta^2v_2\|_2^2).$$

For a function $w\in L^2(\Omega)$ such that $\|w\|_2=1$. Consider the scalar product

$$(w,(u_1-u_2)_t)_2+(w,\Delta^2(u_1-u_2))_2=(w,\det(D^2v_1)-\det(D^2v_2))_2.$$

We have the estimate

$$(w,(u_1-u_2)_t)_2\le\|w\|_2\|\Delta^2(u_1-u_2)\|_2+\|w\|_2\|\det(D^2v_1)-\det(D^2v_2)\|_2,$$

and taking the supremum of all $w\in L^2(\Omega)$ such that $\|w\|_2=1$ we get

$$\sup_w(w,(u_1-u_2)_t)_2\le\|\Delta^2(u_1-u_2)\|_2+\|\det(D^2v_1)-\det(D^2v_2)\|_2.$$

Therefore, from (4.41) we conclude that

$$\|(u_1-u_2)_t\|_2^2 \le C\left[\|\Delta^2(u_1-u_2)\|_2^2+(\|\Delta^2 v_1\|_2^2+\|\Delta^2 v_2\|_2^2)\|\Delta(v_1-v_2)\|_2^2\right],$$

and consequently, by using (4.42),

$$\max_{0\le t\le T}\|\Delta(u_1-u_2)\|_2^2+\int_0^T\|\Delta^2(u_1-u_2)\|_2^2+\int_0^T\|(u_1-u_2)_t\|_2^2 \le C\max_{0\le t\le T}\|\Delta(v_1-v_2)\|_2^2\int_0^T(\|\Delta^2 v_1\|_2^2+\|\Delta^2 v_2\|_2^2). \tag{4.43}$$

On the space $\mathcal{X}_T$ we define the norm

$$\|u\|_{\mathcal{X}_T}^2 := \max_{0\le t\le T}\|\Delta u\|_2^2+\int_0^T\|\Delta^2 u\|_2^2+\int_0^T\|u_t\|_2^2,$$

so that (4.43) reads

$$\|u_1-u_2\|_{\mathcal{X}_T}\le C\left[\int_0^T(\|\Delta^2 v_1\|_2^2+\|\Delta^2 v_2\|_2^2)\right]^{1/2}\|v_1-v_2\|_{\mathcal{X}_T}. \tag{4.44}$$

Now consider the unique solution u_ℓ (see Theorem 4.21) to the linear problem

$$(u_\ell)_t+\Delta^2 u_\ell=\lambda f,$$

with the same boundary and initial conditions as (4.39). Consider the ball

$$B_\rho=\{u\in\mathcal{X}_T : \|u-u_\ell\|_{\mathcal{X}_T}\le\rho\}. \tag{4.45}$$

Using estimate (4.44) we find

$$\|u_i-u_\ell\|_{\mathcal{X}_T}\le C\left(\int_0^T\|\Delta^2 v_i\|_2^2\right)^{1/2}\|v_i\|_{\mathcal{X}_T}\le C\|v_i\|_{\mathcal{X}_T}^2, \tag{4.46}$$

for $i=1,2$. Therefore,

$$\|v_i\|_{\mathcal{X}_T}\le\|v_i-u_\ell\|_{\mathcal{X}_T}+\|u_\ell\|_{\mathcal{X}_T} \tag{4.47}$$

and taking into account that (see Theorem 4.21)

$$\|u_\ell\|_{\mathcal{X}_T}^2\le C\left(\|\Delta u_0\|_2^2+\lambda^2\int_0^T\|f\|_2^2\right)=: C\,\Gamma(\rho,u_0,\lambda,f). \tag{4.48}$$

from (4.46)–(4.47)–(4.48) we obtain that

$$\|u_i - u_\ell\|_{\mathcal{X}_T} \le C\left(\rho^2 + \|\Delta u_0\|_2^2 + \lambda^2 \int_0^T \|f\|_2^2\right),$$

and thus

$$\|u_i - u_\ell\|_{\mathcal{X}_T} \le \rho,$$

for small enough ρ, $|\lambda|$ and $\|\Delta u_0\|_2$.

By using (4.46)–(4.47)–(4.48) and reasoning as before we can transform (4.44) into

$$\|u_1 - u_2\|_{\mathcal{X}_T} \le C\Gamma(\rho, u_0, \lambda, f)^{1/2}\|v_1 - v_2\|_{\mathcal{X}_T}.$$

Again, for ρ, $|\lambda|$ and $\|\Delta u_0\|_2$ small enough we have

$$\|u_1 - u_2\|_{\mathcal{X}_T} \le \frac{1}{2}\|v_1 - v_2\|_{\mathcal{X}_T}.$$

The existence of a unique solution follows from the application of Banach fixed point theorem to the map

$$\begin{aligned} \mathcal{A} : B_\rho &\to B_\rho \\ v_i &\mapsto u_i, \end{aligned}$$

for $i = 1, 2$. The case $T = \infty$ follows similarly since $\Gamma(\rho, u_0, \lambda, f)$ is independent of T.

Step 2. Local Existence in Time By the Gagliardo-Nirenberg inequality [76, 119],

$$\|\Delta v_i\|_\infty \le C\|\Delta v_i\|_2^{1/4}\|\nabla\Delta v_i\|_3^{3/4}, \qquad (i = 1, 2),$$

(4.40) and (4.41) give

$$\begin{aligned} &\|\det(D^2 v_1) - \det(D^2 v_2)\|_2^2 \le C(\|\Delta v_1\|_\infty^2 + \|\Delta v_2\|_\infty^2)\|v_1 - v_2\|^2 \le \\ &C(\|\Delta v_1\|_2^{1/2}\|\nabla\Delta v_1\|_3^{3/2} + \|\Delta v_2\|_2^{1/2}\|\nabla\Delta v_2\|_3^{3/2})\|v_1 - v_2\|^2. \end{aligned}$$

Then by the Sobolev embedding, we find

$$\begin{aligned} &\frac{d}{dt}\|u_1 - u_2\|^2 + \|\Delta^2(u_1 - u_2)\|_2^2 \\ &\quad \le C(\|\Delta v_1\|_2^{1/2}\|\Delta^2 v_1\|_2^{3/2} + \|\Delta v_2\|_2^{1/2}\|\Delta^2 v_2\|_2^{3/2})\|v_1 - v_2\|^2. \end{aligned}$$

Integrating respect to time we obtain

$$\max_{0\le t\le T} \|u_1 - u_2\|^2 + \int_0^T \|\Delta^2(u_1-u_2)\|_2^2 \le C \max_{0\le t\le T} \|v_1 - v_2\|^2$$
$$\times C\left(\max_{0\le t\le T} \|v_1\|^{1/2} \int_0^T \|\Delta^2 v_1\|_2^{3/2} + \max_{0\le t\le T} \|v_2\|^{1/2} \int_0^T \|\Delta^2 v_2\|_2^{3/2}\right).$$

By Hölder inequality we get

$$\max_{0\le t\le T} \|u_1 - u_2\|^2 + \int_0^T \|\Delta^2(u_1-u_2)\|_2^2 \le C T^{1/4} \max_{0\le t\le T} \|v_1 - v_2\|^2$$
$$\times\left[\max_{0\le t\le T} \|v_1\|^{1/2} \left(\int_0^T \|\Delta^2 v_1\|_2^2\right)^{3/4} + \max_{0\le t\le T} \|v_2\|^{1/2} \left(\int_0^T \|\Delta^2 v_2\|_2^2\right)^{3/4}\right].$$

Combining the estimates above with the arguments in Step 1 yields

$$\|u_1 - u_2\|_{\mathcal{X}_T} \le C T^{1/4} \|v_1 - v_2\|_{\mathcal{X}_T}$$
$$\times\left[\max_{0\le t\le T} \|v_1\|^{1/2} \left(\int_0^T \|\Delta^2 v_1\|_2^2\right)^{3/4} + \max_{0\le t\le T} \|v_2\|^{1/2} \left(\int_0^T \|\Delta^2 v_2\|_2^2\right)^{3/4}\right]^{1/2}.$$

Consider again the ball B_ρ defined in (4.45). In this case we have

$$\|u_i - u_\ell\|_{\mathcal{X}_T} \le C T^{1/4} \max_{0\le t\le T} \|v_i\|^{1/4} \left(\int_0^T \|\Delta^2 v_i\|_2^2\right)^{3/8} \|v_i\|_{\mathcal{X}_T}$$
$$\le C T^{1/4} \|v_i\|_{\mathcal{X}_T}^2,$$

for $i = 1, 2$. Arguing as in Step 1 of the present proof we get

$$\|u_i - u_\ell\|_{\mathcal{X}_T} \le C T^{1/4} \Gamma(\rho, u_0, \lambda, f),$$

and thus

$$\|u_i - u_\ell\|_{\mathcal{X}_T} \le \rho,$$

for small enough T. Additionally we have

$$\|u_1 - u_2\|_{\mathcal{X}_T} \le C T^{1/4} \Gamma(\rho, u_0, \lambda, f)^{1/2} \|v_1 - v_2\|_{\mathcal{X}_T}.$$

Again, for T small enough we find

$$\|u_1 - u_2\|_{\mathcal{X}_T} \le \frac{1}{2} \|v_1 - v_2\|_{\mathcal{X}_T}.$$

The existence of a unique solution to (4.38) follows from the application of Banach fixed point theorem to the map

$$\mathcal{A} : B_\sigma \to B_\sigma$$
$$v_i \mapsto u_i \qquad (i = 1, 2).$$

We have so found $\overline{T} = \overline{T}(\lambda, \|u_0\|)$ such that (4.38) admits a unique solution over $[0, T]$ for all $T < \overline{T}$.

Step 3. Blow-Up We argue by contradiction. Assume $[0, T^*)$, with $T^* < \infty$, is the maximal interval of continuation of the solution, and that $\liminf_{t\to T^*} \|u(t)\| = \gamma < \infty$. Then there exists a sequence $\{t_n\}$ such that $t_n \to T^*$ and $\|u(t_n)\| < 2\gamma$ for n large enough. Take n sufficiently large so that $t_n + \overline{T}(\lambda, 2\gamma) > T^*$, where $\overline{T}$ is defined at the end of Step 2. Consider $u(t_n)$ as initial condition to (4.38). Then Step 2 tells us that the solution may be continued beyond T^*, contradiction. □

Corollary 4.23 *Let u be a solution as described in Theorem 4.22 during the time interval* $(0, T]$. *Then there exists a real number $\epsilon > 0$ such that the solution can be prolonged to the interval* $(0, T + \epsilon]$.

Proof This result is a consequence of Step 3 in the proof of Theorem 4.22. □

If $\lambda = 0$ the solution has a higher regularity.

Corollary 4.24 *Let u be a solution as described in Theorem 4.22 to Eq.* (4.36) *with* $\lambda = 0$. *Then* $u^2 \in C^1(0, T; L^1(\Omega))$.

Proof The regularity proven in Theorem 4.22 for the solution u to (4.36) implies that $\det(D^2u) \in C(0, T; L^1(\Omega))$ and $\Delta^2 u \in C(0, T; W^{-2,2}(\Omega))$ so that $u_t \in C(0, T; W^{-2,2}(\Omega))$ and, in turn, $u \in C^1(0, T; W^{-2,2}(\Omega))$. Combined with $u \in C(0, T; W_0^{2,2}(\Omega))$ this yields $uu_t \in C(0, T; L^1(\Omega))$ and, additionally, $u^2 \in C^1(0, T; L^1(\Omega))$. □

4.3.1 Some Results on Asymptotic Behavior

We concentrate our attention in the parabolic problem with $\lambda = 0$, that is

$$\begin{cases} u_t + \Delta^2 u = \det(D^2u) & (x, t) \in \Omega \times (0, T), \\ u(x, 0) = u_0(x) & x \in \Omega \\ u = u_\nu = 0 & (x, t) \in \partial\Omega \times (0, T). \end{cases} \tag{4.49}$$

The goal in this section is to find sufficient conditions for the global existence and the finite blow-up of the solutions.

Preliminary Results We start with the following result that takes advantage of the variational structure of the problem.

Lemma 4.25 *If $u = u(t)$ solves* (4.49) *then its energy*

$$J(u(t)) = \frac{1}{2}\int_\Omega |\Delta u(t)|^2 - \int_\Omega u_x(t)u_y(t)u_{xy}(t)$$

satisfies

$$\frac{d}{dt}J(u(t)) = -\int_\Omega u_t(t)^2 \leq 0\,.$$

Proof It is a direct computation. See [67] for the details. □

Next we try to study the behavior in time of the energy functional, according with the position of a regular data ($u_0 \in W_0^{2,2}(\Omega)$) respect to the Nehari manifold.

Lemma 4.26 *Let $u_0 \in W_0^{2,2}(\Omega)$ be such that $J(u_0) < d$. Then:*

(i) if $u_0 \in \mathcal{N}_-$ the solution $u = u(t)$ to (4.49) *satisfies $J(u(t)) < d$ and $u(t) \in \mathcal{N}_-$ for all $t \in (0, T)$;*
(ii) if $u_0 \in \mathcal{N}_+$ the solution $u = u(t)$ to (4.49) *satisfies $J(u(t)) < d$ and $u(t) \in \mathcal{N}_+$ for all $t \in (0, T)$.*

Proof If $J(u_0) < d$, then $J(u(t)) < d$ for all $t \in (0, T)$ in view of Lemma 4.25. Assume moreover that $u_0 \in \mathcal{N}_+$ and, for contradiction, that $u(t) \notin \mathcal{N}_+$ for some $t \in (0, T)$. Then, necessarily $u(t) \in \mathcal{N}$ for some $t \in (0, T)$ so that, by (4.28), $J(u(t)) \geq d$, contradiction. We may argue similarly if $u_0 \in \mathcal{N}_-$. □

The following result proves a kind of L^2-*Cauchy property* for global solutions with bounded energy.

Lemma 4.27 *Let $u_0 \in W_0^{2,2}(\Omega)$ and let $u = u(t)$ be the corresponding solution to* (4.49). *Then*

$$\|u(t+\delta) - u(t)\|_2^2 \leq \delta\Big(J(u(t)) - J(u(t+\delta))\Big) \qquad \forall\delta > 0$$

and

$$\left(\frac{\|u(t+\delta)\|_2 - \|u(t)\|_2}{\delta}\right)^2 \leq \frac{J(u(t)) - J(u(t+\delta))}{\delta}\,. \tag{4.50}$$

In particular, the map $t \mapsto \|u(t)\|_2$ is differentiable and

$$\left(\frac{d}{dt}\|u(t)\|_2\right)^2 \leq -\frac{d}{dt}J(u(t))\,.$$

Proof By Hölder inequality, Fubini Theorem, and Lemma 4.25, we get

$$\begin{aligned}\|u(t+\delta)-u(t)\|_2^2 &= \int_\Omega \left|\int_t^{t+\delta} u_t(\tau)\right|^2 \le \delta \int_\Omega \int_t^{t+\delta} u_t(\tau)^2 \\ &= \delta \int_t^{t+\delta} \left(\int_\Omega u_t(\tau)^2\right) = \delta\Big(J(u(t)) - J(u(t+\delta))\Big)\end{aligned}$$

which is the first inequality. By the triangle inequality and the just proved inequality we infer that

$$\Big(\|u(s+\delta)\|_2 - \|u(s)\|_2\Big)^2 \le \|u(s+\delta)-u(s)\|_2^2 \le \delta\Big(J(u(t)) - J(u(t+\delta))\Big)$$
$$\forall \delta > 0$$

which we may rewrite as (4.50). Finally, the estimate of the derivative follows by letting $\delta \to 0$. □

Also we have the following identity.

Lemma 4.28 *Let* $u_0 \in W_0^{2,2}(\Omega)$ *and let* $u = u(t)$ *be the corresponding solution to* (4.49). *Then for all* $t \in [0, T)$ *we have*

$$\frac{1}{2}\frac{d}{dt}\|u(t)\|_2^2 + \|u(t)\|^2 - 3\int_\Omega u_x(t)u_y(t)u_{xy}(t) = 0\,. \tag{4.51}$$

Proof Multiply (4.49) by $u(t)$, integrate over Ω, and apply (4.8) to obtain (4.51). □

Finally, we prove that the nonlinear terms goes to the *correct* limit for $W_0^{2,2}(\Omega)$-bounded sequences. We again need to use the cancelation properties of the determinant in the framework of the compensation-compactness.

Lemma 4.29 *Let* $\{u_k\}$ *be a bounded sequence in* $W_0^{2,2}(\Omega)$. *Then there exists* $\overline{u} \in W_0^{2,2}(\Omega)$ *such that* $u_k \rightharpoonup \overline{u}$ *in* $W_0^{2,2}(\Omega)$ *and*

$$\int_\Omega \phi \ \det(D^2 u_k) \to \int_\Omega \phi \ \det(D^2 \overline{u}) \quad \forall \phi \in W_0^{2,2}(\Omega),$$

after passing to a suitable subsequence.

Proof The first part is immediate and follows from the reflexivity of the Sobolev space $W_0^{2,2}(\Omega)$. The second part cannot be deduced in the same way because $L^1(\Omega)$ is not reflexive and consequently the sequence $\det(D^2 u_k)$ could converge to a measure. For all $v, w \in C_0^\infty(\Omega)$ some integrations by parts show that

$$\int_\Omega w \det\left(D^2 v\right) = \int_\Omega v_{x_1} v_{x_2} w_{x_1x_2} - \frac{1}{2} v_{x_2}^2 w_{x_1x_1} - \frac{1}{2} v_{x_1}^2 w_{x_2x_2}. \tag{4.52}$$

A density argument shows that the same is true for all $v, w \in W_0^{2,2}(\Omega)$. Therefore for any $\phi \in W_0^{2,2}(\Omega)$ and any k we have

$$\int_\Omega \phi \det\left(D^2 u_k\right) = \int_\Omega (u_k)_{x_1} (u_k)_{x_2} \phi_{x_1x_2} - \frac{1}{2} (u_k)_{x_2}^2 \phi_{x_1x_1} - \frac{1}{2} (u_k)_{x_1}^2 \phi_{x_2x_2}.$$

By compact embedding we know that $u_k \to u$ strongly in $W_0^{1,4}(\Omega)$ since $u_k \rightharpoonup u$ weakly in $W_0^{2,2}(\Omega)$, and thus

$$\lim_{k\to\infty} \int_\Omega \phi \det\left(D^2 u_k\right) = \int_\Omega \overline{u}_{x_1} \overline{u}_{x_2} \phi_{x_1x_2} - \frac{1}{2} \overline{u}_{x_2}^2 \phi_{x_1x_1} - \frac{1}{2} \overline{u}_{x_1}^2 \phi_{x_2x_2},$$

after passing to a suitable subsequence. Applying again (4.52) leads to

$$\lim_{k\to\infty} \int_\Omega \phi \det\left(D^2 u_k\right) = \int_\Omega \phi \det\left(D^2 \overline{u}\right),$$

up to a subsequence. □

Finite Time Blow-Up The order of the equation motivates the use of energy estimates to analyze the blow-up. The arguments by comparison, in general, do not hold for order greater that two.

We give two types of results related to the blow-up in finite time.

(1) Results Related to the Mountain-Pass Level Our first result proves the existence of solutions to (4.49) which blow up in finite time, related to the *mountain-pass* level of the energy functional J.

Theorem 4.30 *Let $u_0 \in \mathcal{N}_-$ be such that $J(u_0) \le d$. Then the solution $u = u(t)$ to* (4.49) *blows up in finite time, that is, there exists $T > 0$ such that $\|u(t)\| \to +\infty$ as $t \nearrow T$. Moreover, the blow up also occurs in the $W_0^{1,4}(\Omega)$-norm, that is, $\|u(t)\|_{W_0^{1,4}(\Omega)} \to +\infty$ as $t \nearrow T$.*

Proof Since $u_0 \notin \mathcal{N}$, we know that, by Lemma 4.25, we have $J(u(t)) < d$ for all $t > 0$. Therefore, possibly by translating t, we may assume that $J(u(0)) < d$ and, from now on, we rename $u_0 = u(0)$. We use here a refinement of the concavity method by Levine [103], see also [125, 144].

Assume for contradiction that the solution $u = u(t)$ to (4.49) is global and define

$$M(t) := \frac{1}{2} \int_0^t \|u(s)\|_2^2,$$

by Theorem 4.22 and Corollary 4.24, $M \in C^2(0,\infty)$. Then by direct calculation we find that

$$M'(t) = \frac{\|u(t)\|_2^2}{2}$$

and

$$M''(t) = -3J(u(t)) + \frac{\|u(t)\|^2}{2}$$

also by using (4.51). By the assumptions on u_0 and by Lemma 4.26 we know that $u(t) \in N_-$ for all $t \geq 0$. Hence, by Theorem 4.14, we conclude that $\|u(t)\|^2 > 6d$ for all $t \geq 0$. Therefore by Lemma 4.25 and the assumptions, we get

$$M''(t) \geq -3J(u_0) + \frac{\|u(t)\|^2}{2} > 3(d - J(u_0)) > 0 \qquad \text{for all } t \geq 0\,.$$

Then

$$\lim_{t\to\infty} M(t) = \lim_{t\to\infty} M'(t) = +\infty\,. \tag{4.53}$$

By Lemma 4.25 we also have that

$$J(u(t)) = J(u_0) - \int_0^t \|u_t(s)\|_2^2$$

so that

$$M''(t) = 3\int_0^t \|u_t(s)\|_2^2 - 3J(u_0) + \frac{\|u(t)\|^2}{2} > 3\int_0^t \|u_t(s)\|_2^2$$

since $\|u(t)\|^2 > 6d > 6J(u_0)$. By multiplying the previous inequality by $M(t) > 0$ and by using Hölder inequality, we get

$$M''(t)M(t) \geq \frac{3}{2}\int_0^t \|u_t(s)\|_2^2 \int_0^t \|u(s)\|_2^2 \geq$$
$$\frac{3}{2}\left(\int_0^t \int_\Omega u(s)u_t(s)\right)^2 = \frac{3}{2}\Big(M'(t) - M'(0)\Big)^2 .$$

By (4.53) we know that there exists $\tau > 0$ such that $M'(t) > 7M'(0)$ for $t > \tau$ so that the latter inequality becomes

$$M''(t)M(t) > \frac{54}{49}M'(t)^2 \qquad \text{for all } t > \tau\,. \tag{4.54}$$

This shows that the map $t \mapsto M(t)^{-5/49}$ has negative second derivative and is therefore concave on $[\tau, +\infty)$. Since $M(t)^{-5/49} \to 0$ as $t \to \infty$ in view of (4.53), we reach a contradiction. This shows that the solution $u(t)$ is not global and, by Theorem 4.22, that there exists $T > 0$ such that $\|u(t)\| \to +\infty$ as $t \nearrow T$.

Since by Lemma 4.26 we have that $u(t) \in \mathcal{N}_-$ for all $t \geq 0$, by (3.17) we find that

$$\|u(t)\|^2 < 3 \int_\Omega u_x(t)u_y(t)u_{xy}(t) \leq \frac{3}{4}\|u(t)\| \, \|u(t)\|^2_{W_0^{1,4}(\Omega)} \qquad \text{for all } t \geq 0$$

so that $\|u(t)\| < \frac{3}{4}\|u(t)\|^2_{W_0^{1,4}(\Omega)}$ and the $W_0^{1,4}(\Omega)$-norm also blows up as $t \nearrow T$. □

(2) Results of Blow-Up Independent of the Mountain-Pass Level Next, we state a blow up result without assuming that the initial energy $J(u_0)$ is smaller than the mountain pass level d.

Let λ_1 denote the least Dirichlet eigenvalue of the biharmonic operator in Ω and assume that $u_0 \in W_0^{2,2}(\Omega)$ satisfies

$$\lambda_1 \|u_0\|_2^2 > 6J(u_0) \, . \tag{4.55}$$

By Poincaré inequality $\|u_0\|^2 \geq \lambda_1 \|u_0\|_2^2$, we see that if u_0 satisfies (4.55), then $u_0 \in \mathcal{N}_-$. However, the energy $J(u_0)$ may be larger than d.

Indeed, let e^1 denote an eigenfunction corresponding to λ_1 with the sign implying $\int_\Omega e^1_x e^1_y e^1_{xy} > 0$. If we take $u_0 = \alpha e^1$, then (4.55) will be satisfied for any $\alpha > \overline{\alpha}$ where $\overline{\alpha}$ is the unique value of $\alpha > 0$ such that $\overline{\alpha} e^1 \in \mathcal{N}$. And, by (4.28), we know that $J(\overline{\alpha} e^1) > d$. So, for $\alpha > \overline{\alpha}$ sufficiently close to $\overline{\alpha}$ we have $J(\alpha e^1) > d$, that is, we are above the mountain pass level.

As a previous step to the blow-up result we prove the following proposition that is one of the peculiarities of the behavior of the fourth order problem, that is, if a solution is defined for all t then the norm is globally bounded. In others words there is not blow-up at infinity. This behavior is new respect, for instance, to semilinear second order parabolic equations at critical growth, see [118, 132].

Proposition 4.31 *Assume that $u = u(t)$ is a global solution to* (4.49), *then*

$$\liminf_{t \to \infty} \|u(t)\| < +\infty \, . \tag{4.56}$$

Proof If, by contradiction, the solution $u = u(t)$ to (4.49) is global and

$$\|u(t)\| \to +\infty \qquad \text{as } t \to +\infty \, , \tag{4.57}$$

consider

$$M(t) := \frac{1}{2} \int_0^t \|u(s)\|_2^2$$

as in the proof of Theorem 4.30. Then

$$M''(t) = -3J(u(t)) + \frac{\|u(t)\|^2}{2} \to +\infty \qquad \text{as } t \to +\infty$$

because the map $t \mapsto -3J(u(t))$ is increasing, according to (4.57) and Lemma 4.25. By Lemma 4.25 and using (4.57), there exists $\tau > 0$ such that

$$M''(t) > 3\int_0^t \|u_t(s)\|_2^2 \qquad \forall t > \tau\,.$$

By multiplying the previous inequality by $M(t) > 0$ and by using Hölder inequality, we find

$$M''(t)M(t) \geq \frac{3}{2}\Big(M'(t) - M'(0)\Big)^2 \qquad \forall t > \tau$$

and that (4.54) holds, for a possibly larger τ. By the same concavity argument as in the proof of Theorem 4.30, we reach a contradiction. As a consequence (4.57) cannot occur and (4.56) follows. □

Next we prove the main result of this part.

Theorem 4.32 *Assume that* $u_0 \in W_0^{2,2}(\Omega)$ *satisfies* (4.55). *Then the solution* $u = u(t)$ *to* (4.49) *blows up in finite time, that is, there exists* $T > 0$ *such that* $\|u(t)\| \to +\infty$ *and* $\|u(t)\|_{W_0^{1,4}(\Omega)} \to +\infty$ *as* $t \nearrow T$.

Proof By using the Poincaré inequality and Lemma 4.25, (4.51) yields

$$\frac{d}{dt}\|u(t)\|_2^2 = -6J(u(t)) + \|u(t)\|^2 \geq -6J(u_0) + \lambda_1\|u(t)\|_2^2\,.$$

By putting $\psi_0(t) := -6J(u_0) + \lambda_1\|u(t)\|_2^2$, the previous inequality reads $\psi_0'(t) \geq \lambda_1\psi_0(t)$. Since (4.55) yields $\psi_0(0) > 0$, this proves that $\psi_0(t) \to \infty$ as $t \to \infty$. Hence, by invoking again Poincaré inequality, we see that also (4.57) holds, a situation that we eliminate by proving (4.56). This contradiction shows that $T < \infty$.

The blow up of the $W_0^{1,4}(\Omega)$-norm follows as in the proof of Theorem 4.30. □

Let $u_0 \in W_0^{2,2}(\Omega)$ and let $u = u(t)$ be the local solution to (4.49). According to Theorem 4.22, the solution blows up at some $T > 0$ if

$$\lim_{t \to T} \|u(t)\| = +\infty\,. \tag{4.58}$$

We are able to prove the following forms of blow up:

$$\lim_{t\to T} \|u\|_{L^2((0,t);W_0^{2,2}(\Omega))} = +\infty\,, \tag{4.59}$$

$$\lim_{t\to T} \|u(t)\|_2 = +\infty\,, \tag{4.60}$$

$$\lim_{t\to T} \|u\|_{L^4((0,t);W_0^{1,4}(\Omega))} = +\infty\,. \tag{4.61}$$

Clearly, (4.60) implies (4.58). Moreover further implications hold true.

Theorem 4.33 *Let $u_0 \in W_0^{2,2}(\Omega)$ and let $u = u(t)$ be the local solution to* (4.49). *Assume that* (4.58) *occurs for some finite $T > 0$. Then there exists $\tau \in (0, T)$ such that $u(t) \in \mathcal{N}_-$ for all $t > \tau$.*

Moreover:

(i) If (4.59) *occurs, then* (4.60) *occurs.*
(ii) If (4.60) *occurs, then* (4.61) *occurs.*

Finally, (4.60) *occurs if and only if*

$$\lim_{t\to T} \int_0^t \left(\int_\Omega \Delta u(s)|\nabla u(s)|^2\right) = -\infty\,. \tag{4.62}$$

See the details in [67].

Global Solutions For suitable initial data, not only the solution is global but it vanishes in infinite time, that is, we find some information about the basin of attraction of the trivial stationary solution. More precisely we have the following result.

Theorem 4.34 *Let $u_0 \in \mathcal{N}_+$ be such that $J(u_0) \le d$. Then the solution $u = u(t)$ to* (4.49) *is global and $u(t) \to 0$ in $W^{4,2}(\Omega)$ as $t \to +\infty$.*

Proof Since $u_0 \notin \mathcal{N}$, we know that it is not a stationary solution to (4.49), that is, it does not solve (4.23). Hence, by Lemma 4.25 we have $J(u(t)) < d$ for all $t > 0$. By Lemma 4.26 and Theorem 4.14 we know that $u(t)$ remains bounded in $W_0^{2,2}(\Omega)$ so that, by Theorem 4.30, the solution is global. If $\|u_t\|_2 \ge c > 0$ for all $t > 0$, then by Lemma 4.25 we would get $J(u(t)) \to -\infty$ as $t \to \infty$ against $u(t) \in N_+$, see again Lemma 4.26. Hence, $u_t(t) \to 0$ in $L^2(\Omega)$, on a suitable sequence.

Moreover, the boundedness of $\|u(t)\|$ implies that there exists $\overline{u} \in W_0^{2,2}(\Omega)$ such that $u(t) \rightharpoonup \overline{u}$ in $W_0^{2,2}(\Omega)$ as $t \to \infty$ on the sequence. Note also that, by Lemma 4.29, for all $\phi \in W_0^{2,2}(\Omega)$ we have

$$\int_\Omega \phi \det(D^2u(t)) \to \int_\Omega \phi \det(D^2\overline{u}).$$

Therefore, if we test (4.49) with some $\phi \in W_0^{2,2}(\Omega)$, and we let $t \to \infty$ on the above found sequence, we get

$$0 = \int_\Omega u_t(t)\phi + \int_\Omega \Delta u(t)\Delta\phi - \int_\Omega \det(D^2u(t))\phi \to \int_\Omega \Delta\overline{u}\Delta\phi - \det(D^2\overline{u})\phi$$

which shows that $\overline{u}$ solves (4.23). Since the only solution to (4.23) at energy level below d is the trivial one, we infer that $\overline{u} = 0$. Writing (4.49) as

$$\Delta^2 u(t) = -u_t(t) + \det(D^2u(t))$$

we see that $\Delta^2 u(t)$ is uniformly bounded in $L^1(\Omega)$.

With similar arguments an in the proof of Theorem 4.12, we see that $\Delta^2 u(t)$ is bounded in $W^{-s,2}(\Omega)$ for all $s > 1$ and, by a bootstrap argument, that

$$\Delta^2 u(t) = -u_t(t) + \det(D^2u(t)) \to 0 \text{ strongly in } L^2(\Omega)$$

so that $u(t) \to 0$ in $W^{4,2}(\Omega)$ on the same sequence. We infer by Lemma 4.25 that $J(u(t)) \to 0$ as $t \to \infty$. Since $u(t) \in \mathcal{N}_+$ for all $t \geq 0$, we also have that $J(u(t)) \geq \|u(t)\|^2/6$ for all t.

Therefore we conclude that all the above convergence occur as $t \to \infty$, not only on some subsequence. □

Description of the ω-Limit Set Theorems 4.30 and 4.32 determine a wide class of initial data $u_0 \in W_0^{2,2}(\Omega)$ which ensure that the solution to (4.49) blows up in finite time. We have shown in Proposition 4.31 that infinite time blow up cannot occur for the fourth order parabolic equation (4.49). If $T = +\infty$, we denote by

$$\omega(u_0) = \bigcap_{t\geq 0} \overline{\{u(s) \,:\, s \geq t\}}$$

the ω-limit set of $u_0 \in W_0^{2,2}(\Omega)$, where the closure is taken in $W_0^{2,2}(\Omega)$.

Remark 4.35 An important fact must be considered: Since the nonlinearity appearing in our problem, $det(D^2u)$, is analytic, then for any bounded trajectory the ω-limit set is only one point. See [84, 86] and the references therein.

The following sharper result can be also found in [67].

Theorem 4.36 *Let $u_0 \in W_0^{2,2}(\Omega)$ and let $u = u(t)$ be the solution to (4.49), such that is defined $(0, +\infty)$. Then the ω-limit set $\omega(u_0)$ is a nonempty bounded connected subset of $W_0^{2,2}(\Omega)$ which consists of solutions to (4.23). In particular, this means that there exists a solution $\overline{u}$ to (4.23) such that $u(t) \to \overline{u}$ in $W_0^{2,2}(\Omega)$ up to a subsequence and, if $\overline{u}$ is an isolated solution to (4.23), then $u(t) \to \overline{u}$ in $W_0^{2,2}(\Omega)$ as $t \to \infty$ (without passing on a subsequence). These convergences are, in fact, also in $W^{4,2}(\Omega)$.*

Notice that in Theorem 4.36, by $\overline{u}$ *is an isolated solution*, we mean that there exists a $W_0^{2,2}(\Omega)$-neighborhood of $\overline{u}$ which contains no further solutions to (4.23). In general, Theorem 4.36 cannot be improved with the statement that the *whole* trajectory converges, see [132] and references therein for second order equations. Note also that from Lemma 4.27 we have that if u is a global solution, then

$$\lim_{t\to\infty} \|u(t+\delta) - u(t)\|_2 = 0 \qquad \forall \delta > 0.$$

This shows that the convergence to $\omega(u_0)$ occurs "slowly".

4.4 Appendix: The Compensation-Compactness That We Need

The following particular case of a result by R. Coifman, P.L. Lions, Y. Meyer and S. Semmes (see [53]) gives a distributional sense to the identities above for functions in $W^{2,2}(\mathbb{R}^N)$.

Lemma 4.37 (*R. Coifman, P.L. Lions, Y. Meyer and S. Semmes*) *Let U, V vector fields in $\mathbb{R}^N$ such that ∇U, $\nabla V \in \left[L^2(\mathbb{R}^N)\right]^{N\times N}$ and $div\,(U) = div\,(V) = 0$ in $\mathcal{D}'(\mathbb{R}^N)$. Then*

$$\sum_{i,j=1}^{N} \partial_{ij}(U_i V_j) \in \mathcal{H}^1(\mathbb{R}^N),$$

where U_i and V_j denote the i-th and j-th components of U and V respectively.

The proof of Lemma 4.37 involves some techniques from Harmonic Analysis and an adaptation of the ideas by Luc Tartar on *compensation compactness*. See, for the last subject, the references [141] and [142]. We will use a localization of the following result, which is a Corollary of Lemma 4.37.

Lemma 4.38 *If $u \in W^{2,2}(\mathbb{R}^N)$, $N \geq 2$ then*

$$\sum_{i\neq j} \left(\partial_{ij}(\partial_i u \partial_j u) - \frac{1}{2}\partial_{ii}((\partial_j u)^2) - \frac{1}{2}\partial_{jj}((\partial_i u)^2) \right) \in \mathcal{H}^1(\mathbb{R}^N).$$

Proof For any $1 \leq i < j \leq N$ we define

$$U^{i,j} = (U^{i,j}_{(1)}, U^{i,j}_{(2)}, \ldots, U^{i,j}_{(N)})$$

where $U^{i,j}_{(k)} = 0$, if $k \neq i, j$ and $U^{i,j}_{(i)} = -\partial_j u$, $U^{i,j}_{(j)} = \partial_i u$. In particular

$$\mathrm{div}(U^{i,j}) = 0.$$

Now let us denote $U = U^{i,j}$ and $V = U^{i,j}$. Then $U_iV_i = (\partial_j u)^2$, $U_jV_j = (\partial_i u)^2$, and $U_iV_j = U_jV_i = -\,\partial_j u\partial_i u$, otherwise $U_kV_l = 0$, whenever $k \neq i, j$ or $l \neq i, j$. In particular recalling Lemma 4.37 it results

$$-\frac{1}{2}\sum_{k,l=1}^{N} \partial_{kl}(U_kV_l) = \partial_{ij}(\partial_i u\partial_j u) - \frac{1}{2}\partial_{ii}((\partial_j u)^2) - \frac{1}{2}\partial_{jj}((\partial_i u)^2) \in \mathcal{H}^1(\mathbb{R}^N).$$

In particular, by linearity, we conclude

$$\sum_{i<j}\left(\partial_{ij}(\partial_i u\partial_j u) - \frac{1}{2}\partial_{ii}((\partial_j u)^2) - \frac{1}{2}\partial_{jj}((\partial_i u)^2)\right) \in \mathcal{H}^1(\mathbb{R}^N)$$

□

By using Lemma 4.38, integrating by parts we get, for every $v \in C_0^\infty(\Omega)$,

$$\begin{aligned}\int_\Omega (\Delta^2 u - S_2(D^2u))v &= \int_\Omega \Delta u\Delta v\\ &- \int_\Omega \sum_{1\le i<j\le N}\left((\partial_i u\partial_j u)\partial_{ij}v + \frac{1}{2}\partial_i(\partial_j u)^2\partial_i v + \frac{1}{2}\partial_j(\partial_i u)^2\partial_j v\right)\\ &= \int_\Omega \Delta u\Delta v - \int_\Omega \sum_{1\le i<j\le N}\Big(\partial_i u\partial_j u\partial_{ij}v + \partial_j u\partial_{ij}u\partial_i v + \partial_i u\partial_{ji}u\partial_j v\Big).\end{aligned} \tag{4.63}$$

Hence, if we consider the functional

$$\mathcal{G}(u) = \frac{1}{2}\int_\Omega |\Delta u|^2 - \int_\Omega \sum_{1\le i<j\le N} \partial_{ij}u\partial_i u\partial_j u,$$

defined in $W_0^{2,2}(\Omega)$, then

$$\left.\frac{d\mathcal{G}(u+\epsilon v)}{d\epsilon}\right|_{\epsilon=0} = \int_\Omega \Delta u\Delta v - \int_\Omega \sum_{1\le i<j\le N}\Big(\partial_i u\partial_j u\partial_{ij}v + \partial_j u\partial_{ij}u\partial_i v + \partial_i u\partial_{ji}u\partial_j v\Big).$$

We can summarize the previous calculation in the following result.

Proposition 4.39 *Consider the functional*

$$\mathcal{G}(u) = \frac{1}{2}\int_\Omega |\Delta u|^2 - \int_\Omega \sum_{1\le i<j\le N} \partial_{ij}u\partial_i u\partial_j u, \tag{4.64}$$

with $u \in W_0^{2,2}(\Omega)$. *Then, the critical points of* $\mathcal{G}$ *are weak solutions to the Dirichlet problem* (4.33).

Note For other results and open problem about the parabolic and the elliptic problems, the reader could see [67, 72] and the references therein.

References

1. B. Abdellaoui, I. Peral, Existence and nonexistence results for quasilinear elliptic equations involving the p-Laplacian. Ann. Mat. Pura. Appl. **182**(3), 247–270 (2003)
2. B. Abdellaoui, I. Peral, The equation $-\Delta u - \lambda \frac{u}{|x|^2} = |\nabla u|^p + cf(x)$, the optimal power. Ann. Sc. Norm. Sup. Pisa (5) **6**(1), 159–183 (2007)
3. B. Abdellaoui, E. Colorado, I. Peral, Existence and nonexistence results for a class of linear and semilinear parabolic equations related to Caffarelli-Kohn-Nirenberg inequalities. J. Eur. Math. Soc. **6**, 119–149 (2004)
4. B. Abdellaoui, A. Dall'Aglio, I. Peral, Some remarks on elliptic problems with critical growth in the gradient. J. Differ. Equ. **222**(1), 21–62 (2006)
5. B. Abdellaoui, A. Dall'Aglio, I. Peral, Regularity and nonuniqueness results for parabolic problems arising in some physical models, having natural growth in the gradient. J. Math. Pures Appl. **90**, 242–269 (2008)
6. B. Abdellaoui, A. Dall'Aglio, I. Peral, S. Segura, Global existence for nonlinear parabolic problems with measure data applications to non-uniqueness for parabolic problemswith critical gradient terms. Adv. Nonlinear Stud. **11**, 733–780 (2011)
7. B. Abdellaoui, D. Giachetti , I. Peral, M. Walias, Elliptic problems with nonlinear terms depending on the gradient and singular on the boundary. Nonlinear Anal. **74**, 1355–1371 (2011)
8. B. Abdellaoui, I. Peral, M. Walias, Multiplicity results for porous media and fast diffusion equations with a quadratic gradient term. Contemp. Math. **594**, 37–58 (2013)
9. B. Abdellaoui, I. Peral, M. Walias, Porous media and fast diffusion equations with a gradient term. Trans. Am. Math. Soc. **367**(7), 4757–4791 (2015)
10. S. Agmon, A. Douglis, L. Nirenberg, Estimates near the boundary for solutions of elliptic partial differential equations satisfying general boundary conditions. I. Commun. Pure Appl. Math. **12**, 623–727 (1959)
11. S. Agmon, A. Douglis, L. Nirenberg, Estimates near the boundary for solutions of elliptic partial differential equations satisfying general boundary conditions. II. Commun. Pure Appl. Math. **17**, 35–92 (1964)
12. N.E. Alaa, M. Pierre, Weak solutions of some quasilinear elliptic equations with data measures. SIAM J. Math. Anal. **24**, 23–35 (1993)
13. A. Alama, Semilinear elliptic equations with sublinear indefinite nonlinearities. Adv. Differ. Equ. **4**(6) , 813–842 (1999)
14. H.W. Alt, S. Luckhaus, Quasilinear elliptic-parabolic differential equations. Math. Z. **183**, 311–341 (1983)
15. A. Ambrosetti, P.H. Rabinowitz, Dual variational methods in critical point theory and applications. J. Funct. Anal. **14**, 349–381 (1973)
16. A. Ambrosetti, H. Brezis, G. Cerami, Combined effects of concave and convex nonlinearities in some elliptic problems. J. Funct. Anal. **122**(2), 519–543 (1994)
17. J.P. Aubin, I. Ekeland, *Applied Nonlinear Analysis* (Wiley, New York, 1984)
18. A.-L. Barabási, H.E. Stanley, *Fractal Concepts in Surface Growth* (Cambridge University Press, Cambridge, 1995)

19. P. Baras, M. Pierre, Singularités éliminables pour des équations semi-linéaires. Ann. Inst. Fourier (Grenoble) **34**(1), 185–206 (1984)
20. P. Baras, M. Pierre, Critère d'existence de solutions positives pour des équations semi-linéaires non monotones. Ann. Inst. H. Poincaré Anal. Non Linéaire **2**(3), 185–212 (1985)
21. G.I. Barenblatt, M. Bertsch, A.E. Chertock, V.M. Prostokishin, Self-similar intermediate asymptotics for a degenerate parabolic filtration-absorption equation. Proc. Natl. Acad. Sci. **97**(18), 9844–9848 (2000)
22. A. Ben-Artzi, P. Souplet, F.B. Weissler, The local theory for the viscous Hamilton-Jacobi equations in Lebesgue spaces. J. Math. Pure. Appl. **9**(81), 343–378 (2002)
23. P. Benilan, H. Brezis, Nonlinear problems related to the Thomas-Fermi equation. Dedicated to Philippe Bénilan. J. Evol. Equ. **3**(4), 673–770 (2003)
24. Ph. Bénilan, P. Wittbold, On mild and weak solutions of elliptic-parabolic problems. Adv. Differ. Equ. **1**(6), 1053–1073 (1996)
25. Ph. Bénilan, P. Wittbold, Sur un problème parabolique-elliptique. M2AN Math. Model. Numer. Anal. **33**(1), 121–127 (1999)
26. Ph. Bénilan, M. Crandall, M. Pierre, Solutions of the porous medium equation in $\mathbb{R}^N$ under optimal conditions on initial values. Indiana Univ. Math. J. **33**(1), 51–87 (1984)
27. P. Bénilan, L. Boccardo, T. Gallouët, R. Gariepy, M. Pierre, J.L. Vázquez, An L^1 theory of existence and uniqueness of solutions of nonlinear elliptic equations. Ann. Sc. Norm. Sup. Pisa (4) **22**(2), 240–273 (1995)
28. H. Berestycki, S. Kamin, G. Sivashinsky, Metastability in a flame front evolution equation. Interfaces Free Bound. **3**(4), 361–392 (2001)
29. D. Blanchard, F. Murat, Renormalised solutions of nonlinear parabolic problems with L^1 data: existence and uniqueness. Proc. R. Soc. Edinb. Sect. A **127**(6), 1137–1152 (1997)
30. D. Blanchard, A. Porretta, Nonlinear parabolic equations with natural growth terms and measure initial data. Ann. Sc. Norm. Sup. Pisa cl. **30**(3–4), 583–622 (2001)
31. D. Blanchard, A. Porretta, Stefan problems with nonlinear diffusion and convection. J. Differ. Equ. **210**, 383–428 (2005)
32. D. Blanchard, F. Murat, H. Redwane, Existence and uniqueness of a renormalized solution for a fairly general class of nonlinear parabolic problems. J. Differ. Equ. **177**(2), 331–374 (2001)
33. L. Boccardo, G. Croce, *Elliptic Partial Differential Equations. Existence and Regularity of Distributional Solutions*. De Gruyter Studies in Mathematics, vol. 55 (De Gruyter, Berlin, 2014)
34. L. Boccardo, T. Gallouët, Nonlinear elliptic and parabolic equations involving measure data. J. Funct. Anal. **87**(1), 149–169 (1989)
35. L. Boccardo, F. Murat, J.-P. Puel, Existence des solutions non bornées pour certains équations quasi-linéaires. Port. Math. **41**, 507–534 (1982)
36. L. Boccardo, F. Murat, J.-P. Puel, Existence de solutions faibles pour des équations elliptiques quasi-linéaires à croissance quadratique, in *Nonlinear Partial Differential Equations and Their Applications, Collège de France Seminar, vol. IV*. Research Notes in Mathematics, vol. 84, ed. by J.-L. Lions, H. Brezis (Pitman, London, 1983), pp. 19–73
37. L. Boccardo, F. Murat, J.-P. Puel, Resultats d'existence pour certains problèmes elliptiques quasi-linéaires. Ann. Sc. Norm. Sup. Pisa **11**(2), 213–235 (1984)
38. L. Boccardo, F. Murat, J.-P. Puel, Existence of bounded solutions for nonlinear elliptic unilateral problems. Ann. Mat. Pura Appl. **152**, 183–196 (1988)
39. L. Boccardo, F. Murat, J.-P. Puel, Existence results for some quasilinear parabolic equations. Nonlinear Anal. Theory Methods Appl. **13**, 378–392 (1989)
40. L. Boccardo, F. Murat, J.-P. Puel, L^∞ estimates for some nonlinear elliptic partial differential equations and application to an existence result. SIAM J. Math. Anal. **2**, 326–333 (1992)
41. L. Boccardo, T. Gallouët, L. Orsina, Existence and uniqueness of entropy solutions for nonlinear elliptic equations with measure data. Ann. Inst. H. Poincaré Anal. Non Linéaire **13**(5), 539–551 (1996)
42. L. Boccardo, A. Dall'Aglio, T. Gallouët, L. Orsina, Nonlinear parabolic equations with measure data. J. Funct. Anal. **147**(1), 237–258 (1997)

43. L. Boccardo, S. Segura de León, C. Trombetti, Bounded and unbounded solutions for a class of quasi-linear elliptic problems with a quadratic gradient term. J. Math. Pures Appl. **80**(9), 919–940 (2001)
44. H. Brezis, X. Cabré, Some simple nonlinear PDE's without solution. Boll. Unione Mat. Ital. Sez. B, **8**, 223–262 (1998)
45. H. Brezis, S. Kamin, Sublinear elliptic equations in $\mathbb{R}^N$. Manuscripta Math. **74**, 87–106 (1992)
46. H. Brezis, M. Marcus, I. Shafrir, Extremal functions for Hardy's inequality with weight. J. Funct. Anal. **171**(1), 177–191 (2000)
47. A.P. Calderón, A. Zygmund, On singular integrals. Am. J. Math. **78**, 289–309 (1956)
48. A.P. Calderón, A. Zygmund, Singular integral operators and differential equations. Am. J. Math. **79**, 901–921 (1957)
49. J. Carrillo, Entropy solutions for nonlinear degenerate problems. Arch. Ration. Mech. Anal. **147**, 269–361 (1999)
50. J. Carrillo, P. Wittbold, Uniqueness of renormalized solutions of degenerate elliptic-parabolic problems. J. Differ. Equ. **156**, 93–121 (1999)
51. D.-C. Chang, G. Dafni, E.M. Stein, Hardy spaces, BMO, and boundary value problems for the Laplacian on a smooth domain in $\mathbb{R}^N$. Trans. Am. Math. Soc. **351**(4), 1605–1661 (1999)
52. K. Cho, H.J. Choe, Nonlinear degenerate elliptic partial differential equations with critical growth conditions on the gradient. Proc. Am. Math. Soc. **123**(12), 3789–3796 (1995)
53. R. Coifman, P.L. Lions, Y. Meyer, S. Semmes, Compensated compactness and Hardy spaces. J. Math. Pures Appl. **72**, 247–286 (1993)
54. A. Dall'Aglio, Approximated solutions of equations with L^1-data. Application to the H-convergence of quasi-linear parabolic equations. Ann. Mat. Pura Appl. (4) **170**, 207–240 (1996)
55. A. Dall'Aglio, D. Giachetti, J.-P. Puel, Nonlinear elliptic equations with natural growth in general domains. Ann. Mat. Pura Appl. **181**, 407–426 (2002)
56. A. Dall'Aglio, D. Giachetti, J.-P. Puel, Nonlinear parabolic equations with natural growth in general domains. Boll. Unione Mat. Ital. Sez. B **8**, 653–683 (2005)
57. A. Dall'Aglio, D. Giachetti, C. Leone, S. Segura de León, Quasilinear parabolic equations with degenerate coercivity having a quadratic gradient term. Ann. Inst. H. Poincaré Anal. Non Linéaire **23**(1), 97–126 (2006)
58. A. Dall'Aglio, D. Giachetti, S. Segura de León, Nonlinear parabolic problems with a very general quadratic gradient term. Differ. Integral Equ. **20**(4), 361–396 (2007)
59. A. Dall'Aglio, D. Giachetti, I. Peral, S. Segura de León, Global existence for some slightly super-linear parabolic equations with measure data. J. Math. Anal. Appl. **345**(2), 892–902 (2008)
60. G. Dal Maso, A. Malusa, Some properties of reachable solutions of nonlinear elliptic equations with measure data. Ann. Sc. Norm. Sup. Pisa Cl. Sci. **25**(1–2), 375–396 (1997)
61. G. Dal Maso, F. Murat, L. Orsina, A. Prignet, Renormalized solutions of elliptic equations with general measure data. Ann. Sc. Norm. Sup. Pisa Cl. Sci. **28**(4), 741–808 (1999)
62. J. Droniou, A. Porretta, A. Prignet, Parabolic capacity and soft measures for nonlinear equations. Potential Anal. **19**(2), 99–161 (2003)
63. L.C. Evans, *Partial Differential Equations*. Graduate Studies in Mathematics, vol. 19 (American Mathematical Society, Providence, 1998)
64. I. Ekeland, On the variational principle. J. Math. Anal. Appl. **47**, 324–353 (1974)
65. C. Escudero, Geometric principles of surface growth. Phys. Rev. Lett. **101**, 196102 (2008)
66. C. Escudero, I. Peral, Some fourth order nonlinear elliptic problems related to epitaxial growth. J. Differ. Equ. **254**, 2515–2531 (2013)
67. C. Escudero, F. Gazzola, I. Peral, Global existence versus blow-up results for a fourth order parabolic PDE involving the Hessian. J. Math. Pures Appl. Available online 23 September 2014. https://doi.org/10.1016/j.matpur.2014.09.007
68. L.C. Evans, R.F. Gariepy, *Measure Theory and Fine Properties of Functions*. Studies in Advanced Mathematics (CRC Press, London, 1992)

69. V. Ferone, F. Murat, Quasilinear problems having quadratic growth in the gradient: an existence result when the source term is small, in *Equations aux Dérivées Partielles et Applications*, Gauthier-Villars, Ed. Sci. Méd. (Elsevier, Paris, 1998), pp. 497–515
70. V. Ferone, F. Murat, Nonlinear problems having natural growth in the gradient: an existence result when the source terms are small. Nonlinear Anal. Theory Methods Appl. **42**(7), 1309–1326 (2000)
71. V. Ferone, M.R. Posteraro, J.M. Rakotoson, Nonlinear parabolic problems with critical growth and unbounded data. Indiana Univ. Math. J. **50**(3), 1201–1215 (2001)
72. F. Ferrari, M. Medina, I. Peral, Biharmonic elliptic problems involving the 2-nd Hessian operator. Calc. Var. Partial Differ. Equ. **51**(3–4), 867–886 (2014)
73. N. Fusco, M. Morini, Equilibrium configurations of epitaxially strained elastic films: second order minimality conditions and qualitative properties of solutions. Arch. Ration. Mech. Anal. **203**, 247–327 (2012)
74. I. Fonseca, N. Fusco, G. Leoni, M. Morini, Motion of elastic thin films by anisotropic surface diffusion with curvature regularization. Arch. Ration. Mech. Anal. **205**, 425–466 (2012)
75. M. Fukushima, K.I. Sato, S. Taniguchi, On the closable parts of pre-dirichlet forms and the fine supports of underlying measures. Osaka J. Math. **28**, 517–535 (1991)
76. E. Gagliardo, Proprietà di alcune classi di funzioni in più variabili. Ricerche Mat. **7**, 102–137 (1958)
77. J. García Azorero, I. Peral, Multiplicity of solutions for elliptics problems with critical exponents or with a non-symmetric term. Trans. Am. Math. Soc. **323**(2), 877–895 (1991)
78. F. Gazzola, H. Grunau, G. Sweers, *Polyharmonic Boundary Value Problems. Positivity Preserving and Nonlinear Higher Order Elliptic Equations in Bounded Domains*. Lecture Notes in Mathematics, vol. 1991 (Springer, Berlin, 2010)
79. D. Giachetti, G. Maroscia, Existence results for a classe of porous medium type equations with quadratic gradient term. J. Evol. Equ. **8**, 155–188 (2008)
80. B. Gilding, M. Guedda, R. Kersner, The Cauchy problem for $u_t - \Delta u = |\nabla u|^q$. J. Math. Anal. Appl. **284**, 733–755 (2003)
81. N. Grenon, Existence results for some quasilinear parabolic problems. Ann. Mat. Pura Appl. (4) **165**, 281–313 (1993)
82. N. Grenon, C. Trombetti, Existence results for a class of nonlinear elliptic problems with p-growth in the gradient. Nonlinear Anal. **52**(3), 931–942 (2003)
83. M. Hairer, Solving the KPZ equation. Ann. Math. **178**, 559–664 (2013)
84. J.K. Hale, G. Raugel, Convergence in gradient-like systems with applications to PDE. Z. Angew. Math. Phys. **43**, 63–124 (1992)
85. K. Hansson, V.G. Maz'ya, I.E. Verbitsky, Criteria of solvability for multidimensional Riccati equations. Ark. Mat. **37**, 87–120 (1999)
86. A. Haraux, M.A. Jendoubi, Convergence of bounded weak solutions of the wave equation with dissipation and analytic nonlinearity. Calc. Var. Partial Differ. Equ. **9**, 95–124 (1999)
87. L. Jeanjean, On the existence of bounded Palais-Smale sequences and application to a Landesman-Lazer-type problem set on R^N. Proc. R. Soc. Edinb. Sect. A **129**(4), 787–809 (1999)
88. P.W. Jones, J.L. Journé, On weak convergence in $\mathcal{H}^1(\mathbb{R}^d)$. Proc. Am. Math. Soc. **120**(1), 137–138 (1994)
89. M. Kardar, G. Parisi, Y.C. Zhang, Dynamic scaling of growing interfaces. Phys. Rev. Lett. **56**, 889–892 (1986)
90. J.L. Kazdan, R.J. Kramer, Invariant criteria for existence of solutions to second-order quasilinear elliptic equations. Commun. Pure Appl. Math. **31**(5), 619–645 (1978)
91. J. Kinnuen, P. Lindqvist, Definition and properties of supersolutions to the porous medium equation. J. Reine Angew. Math. **618**, 135–168 (2008)
92. R.V. Kohn, T.S. Lo, A new approach to the continuum modeling of epitaxial growth: slope selection, coarsening, and the role of the uphill current. Phys. D **161**, 237–257 (2002)
93. R.V. Kohn, F. Otto, Upper bound on coarsening rates. Commun. Math. Phys. **229**, 375–395 (2003)

94. R.V. Kohn, X. Yan, Upper bound on the coarsening rate for an epitaxial growth model. Commun. Pure Appl. Math. **LVI**, 1549–1564 (2003)
95. L. Korkut, M. Pašić, D. Žubrinić, Some qualitative properties of solutions of quasilinear elliptic equations and applications. J. Differ. Equ. **170**, 247–280 (2001)
96. O.A. Ladyzhenskaja, N.N. Ural'ceva, *Linear and Quasi-Linear Elliptic Equations* (Academic Press, New York, 1968)
97. O.A. Ladyzenskaja, V.A. Solonnikov, N.N. Ural'ceva, *Linear and Quasi-Linear Equations of Parabolic Type* Translations of Mathematical Monographs, vol. 23 (American Mathematical Society, Providence, 1968)
98. Z.-W. Lai, S. Das Sarma, Kinetic growth with surface relaxation: continuum versus atomistic models. Phys. Rev. Lett. **66**, 2348–2351 (1991)
99. R. Landes, On the existence of weak solutions for quasilinear parabolic boundary value problems. Proc. R. Soc. Edinb. Sect. A **89**, 217–237 (1981)
100. R. Landes, V. Mustonen, On parabolic initial-boundary value problems with critical growth for the gradient. Ann. Inst. H. Poincaré Anal. Non Linéaire **11**, 135–158 (1994)
101. C. Leone, A. Porretta, Entropy solutions for nonlinear elliptic equations in L^1. Nonlinear Anal. Theory Methods Appl. **32**, 325–334 (1998)
102. J. Leray, J.-L. Lions, Quelques résultats de Višik sur les problèmes elliptiques non linéaires par les méthodes de Minty-Browder. Bull. Soc. Math. Fr. **93**, 97–107 (1965)
103. H.A. Levine, Some nonexistence and instability theorems for solutions of formally parabolic equations of the form $Pu_t = -Au + F(u)$. Arch. Ration. Mech. Anal. **51**, 371–386 (1973)
104. P.L. Lions, Generalized solutions of Hamilton-Jacobi equations. Pitman Res. Notes Math. **62** (1982)
105. T. Lukkari, The porous medium equation with measure data. J. Evol. Equ. **10**, 711–729 (2010)
106. T. Lukkari, The fast diffusion equation with measure data. Nonlinear Differ. Equ. Appl. **19**, 329–343 (2012)
107. J. Mal'y, W. Ziemer, *Fine Regularity of Solutions of Elliptic Partial Differential Equations*. Mathematical Surveys and Monographs, vol. 51 (American Mathematical Society, Providence, 1997)
108. Y. Martel, Complete blow up and global behaviour of solutions of $u_t - \Delta u = g(u)$. Ann. Inst. H. Poincaré Anal. Non Linéaire **15**(6), 687–723 (1998)
109. A.N. Milgram, *Supplement II in Partial Differential Equations*. Lectures in Applied Mathematics, vol III, ed. by L. Bers, F. John, M. Schechter (Interscience, New York, 1964), pp. 229–229
110. J. Moser, A Harnack inequality for parabolic differential equations. Commun. Pure Appl. Math. **17**, 101–134 (1964)
111. S. Müller, Det=det. A remark on the distributional determinant. C. R. Acad. Sci. Paris Ser. I **311**, 13–17 (1990)
112. F. Murat, Compacité par compensation. Ann. Sc. Norm. Sup. Pisa Cl. Sci. (4) **5**(3), 489–507 (1978)
113. F. Murat, L'injection du cone positif de H^{-1} dans $W^{-1,q}$ est compacte pour tout $q < 2$. J. Math. Pures Appl. **60**, 309–322 (1981)
114. F. Murat, *Soluciones renormalizadas de EDP elípticas no lineales* (Laboratoire d'Analyse Numérique, Université Paris VI, Paris, 1993, Preprint)
115. F. Murat, A. Porretta, Stability properties, existence, and nonexistence of renormalized solutions for elliptic equations with measure data. Commun. Partial Differ. Equ. **27**(11–12), 2267–2310 (2002)
116. J.D. Murray, *Mathematical Biology. I. An Introduction*. Interdisciplinary Applied Mathematics, vol. 17, 3rd edn. (Springer, New York, 2002)
117. J.D. Murray, *Mathematical Biology. II. Spatial Models and Biomedical Applications*. Interdisciplinary Applied Mathematics, vol. 18, 3rd edn. (Springer, New York, 2003)
118. W.M. Ni, P.E. Sacks, J. Tavantzis, *On the asymptotic behavior of solutions of certain quasilinear parabolic equations*. J. Differ. Equ. **54**, 97–120 (1984)

119. L. Nirenberg, On elliptic partial differential equations. Ann. Sc. Norm. Sup. Pisa **13**, 115–162 (1959)
120. L. Orsina, Solvability of linear and semilinear eigenvalue problems with L^1 data. Rend. Sem. Mat. Univ. Padova **90**, 207–238 (1993)
121. L. Orsina, Elliptic equations with measure data, in Lecture notes of the course *Analisi Superiore* (2012/2013). http://www1.mat.uniroma1.it/people/orsina/
122. L. Orsina, M.M. Porzio, $L^\infty(Q)$-estimate and existence of solutions for some nonlinear parabolic equations. Boll. Unione Mat. Ital. Sez. B **6-B**, 631–647 (1992)
123. L. Oswald, Isolated positive singularities for a non linear heat equation. Houston J. Math. **14**(4), 543–572 (1988)
124. F. Otto, L^1-contractions and uniqueness for quasi-linear elliptic-parabolic equations. J. Differ. Equ. **131**, 20–38 (1996)
125. L.E. Payne, D.H. Sattinger, Saddle points and instability of nonlinear hyperbolic equations. Isr. J. Math. **22**, 273–303 (1975)
126. I. Peral, J.L. Vázquez, On the stability or instability of the singular solution of the semilinear heat equation with exponential reaction term. Arch. Ration. Mech. Anal. **129**, 201–224 (1995)
127. M. Picone, Sui valori eccezionali di un paramtro da cui dipende una equazione differenziale lineare ordinaria del secondo ordine. Ann. Sc. Norm. Pisa. (1) **11**, 1–144 (1910)
128. M. Pierre, Parabolic capacity and Sobolev spaces. SIAM J. Math. Anal. **14**(3), 522–533 (1983)
129. A. Porretta, Nonlinear equations with natural growth terms and measure data, in *2002-Fez Conference on Partial Differential Equations*. Electronic Journal of Differential Equations Conference, vol. 09 (2002), pp. 183–202
130. A. Porretta, S. Segura de Léon, Nonlinear elliptic equations having a gradient term with natural growth. J. Math. Pures Appl. (9) **85**(3), 465–492 (2006)
131. A. Prignet, Existence and uniqueness of *entropy* solutions of parabolic problems with L^1 data. Nonlinear Anal. Theory Methods Appl. **28**(12), 1943–1954 (1997)
132. P. Quittner, P. Souplet, *Superlinear Parabolic Problems. Blow-Up, Global Existence and Steady States*. Birkhäuser Advanced Texts (Basler Lehrbücher, Basel, 2007)
133. P.H. Rabinowitz, Some global results for nonlinear eigenvalue problems. J. Funct. Anal. **7**, 487–513 (1971)
134. V. Radulescu, M. Willem, Linear elliptic systems involving finite Radon measures. Differ. Int. Equ. **16**(2), 221–229 (2003)
135. S. Segura de León, Existence and uniqueness for L^1 data of some elliptic equations with natural growth. Adv. Differ. Equ. **9**, 1377–1408 (2003)
136. J. Serrin, Pathological solutions of elliptic differential equations. Ann. Sc. Norm. Sup. Pisa **18**, 385–387 (1964)
137. J. Simon, Compact sets in the space $L^p(0, T; B)$. Ann. Mat. Pura Appl. **146**, 65–96 (1987)
138. G. Stampacchia, Le problème de Dirichlet pour les équations elliptiques du second ordre à coefficients discontinus. Ann. Inst. Fourier (Grenoble), **15**, 189–258 (1965)
139. E.M. Stein, *Harmonic Analysis: Real-Variable Methods, Orthogonality and Oscillatory Integrals* (Princeton University Press, Princeton, 1993)
140. E.M. Stein, G. Weiss, On the theory of harmonic functions of several variables. I. The theory of H^p-spaces. Acta Math. **103**, 25–62 (1960)
141. L. Tartar, Compensated compactness and applications to partial differential equations, in *Nonlinear Analysis and Mechanics: Heriot-Watt Symposium*. Research Notes in Mathematics 39, vol. IV (Pitman, Boston, 1979), pp. 136–212
142. L. Tartar, Compacité par compensation: résultats et perspectives, in *Nonlinear Partial Differential Equations and Their Applications. Collège de France Seminar*, Paris, 1981/1982. Research Notes in Mathematics 84, vol. IV (Pitman, Boston, 1983), pp. 350–369
143. C. Trombetti, Non-uniformly elliptic equations with natural growth in the gradient. Potential Anal. **18**(4), 391–404 (2003)
144. M. Tsutsumi, On solutions of semilinear differential equations in a Hilbert space. Math. Japon. **17** 173–193 (1972)

145. J.L. Vazquez, *The Porous Media Equation: Mathematical Theory*. Oxford Mathematical Monographs (Clarendon, Oxford, 2006)
146. Z. Wang, M. Willem, Caffarelli-Kohn-Nirenberg inequalities with remainder terms. J. Funct. Anal. **203**(2), 550–568 (2003)
147. M. Winkler, Global solutions in higher dimensions to a fourth order parabolic equation modeling epitaxial thin-film growth. Z. Angew. Math. Phys. **62**(4), 575–608 (2011)

All Functions Are (Locally) s-Harmonic (up to a Small Error)—and Applications

Enrico Valdinoci

2010 Mathematics Subject Classification. 35R11, 34A08, 60G22

1 Introduction

Given $s \in (0, 1)$, we take into account the so-called s-fractional Laplacian

$$(-\Delta)^s u(x) := \int_{\mathbb{R}^n} \frac{2u(x) - u(x+y) - u(x-y)}{|y|^{n+2s}}\, dy. \tag{1.1}$$

In this definition, u is supposed to be a sufficiently smooth function (to make the integral convergent for small y) and with some growth control at infinity (to make the integral convergent for large y). Also, for the sake of simplicity, a normalizing constant is dropped in (1.1). It is also interesting to observe that, by splitting two

E. Valdinoci (✉)
Dipartimento di Matematica, Università degli Studi di Milano, Milan, Italy

Istituto di Matematica Applicata e Tecnologie Informatiche, Pavia, Italy

School of Mathematics and Statistics, University of Melbourne, Parkville, VIC, Australia

School of Mathematics and Statistics, University of Western Australia, Crawley, Perth, WA, Australia
e-mail: enrico@mat.uniroma3.it

A. Farina, E. Valdinoci (eds.), *Partial Differential Equations and Geometric Measure Theory*, Lecture Notes in Mathematics 2211,
https://doi.org/10.1007/978-3-319-74042-3_3

integrals and changing variables, Eq. (1.1) can be written as

$$
\begin{aligned}
(-\Delta)^s u(x) &= \lim_{\rho\searrow 0} \int_{\mathbb{R}^n\setminus B_\rho} \frac{u(x)-u(x+y)}{|y|^{n+2s}}\,dy + \int_{\mathbb{R}^n\setminus B_\rho} \frac{u(x)-u(x-y)}{|y|^{n+2s}}\,dy \\
&= 2\lim_{\rho\searrow 0} \int_{\mathbb{R}^n\setminus B_\rho} \frac{u(x)-u(x+y)}{|y|^{n+2s}}\,dy \\
&= 2\lim_{\rho\searrow 0} \int_{\mathbb{R}^n\setminus B_\rho(x)} \frac{u(x)-u(y)}{|x-y|^{n+2s}}\,dy \\
&=: 2\,\text{P.V.} \int_{\mathbb{R}^n} \frac{u(x)-u(y)}{|x-y|^{n+2s}}\,dy,
\end{aligned}
\tag{1.2}
$$

where the notation "P.V." stands for "in the Cauchy Principal Value Sense" (and the factor 2 will not be relevant for our purposes).

The fractional Laplacian is one of the most widely studied operators in the recent literature, probably in view of its intrinsic beauty (in spite of the first impression that the definition in (1.1) can produce), of the large variety of different problems related to it, and of its great potentials in modeling real-world phenomena in applied sciences.

The setting in (1.1) is clearly related to an "incremental quotient" of u which gets averaged in all the space. Indeed, roughly speaking, Eq. (1.1) combines together several special features related to the classical Laplacian:

1. The classical Laplacian arises from a second order incremental quotient, namely, for a smooth function u and a small increment h, denoting by $\{e_j\}_{j=1,\dots,n}$ the standard Euclidean basis of $\mathbb{R}^n$, it holds that

$$
\begin{aligned}
&2u(x) - u(x+he_j) - u(x-he_j) \\
&\quad= 2u(x) - \left(u(x) + \nabla u(x)\cdot(he_j) + \frac{1}{2} D^2u(x)(he_j)\cdot(he_j) + o(h^2) \right) \\
&\qquad - \left(u(x) + \nabla u(x)\cdot(-he_j) + \frac{1}{2} D^2u(x)(he_j)\cdot(he_j) + o(h^2) \right) \\
&\quad= -h^2\partial^2_{jj}u(x) + o(h^2)
\end{aligned}
$$

and so

$$
\lim_{h\to 0} \frac{2u(x)-u(x+he_j)-u(x-he_j)}{h^2} = -\Delta u(x).
$$

Comparing this with (1.1), we recognize a structure related to incremental quotients in the definition of fractional Laplacian;

2. The classical Laplacian compares the value of a function with its average. Indeed, for a small $\rho > 0$,

$$\begin{aligned}\int_{B_\rho(x)} u(y)\,dy &= \int_{B_\rho} u(x+y)\,dy \\ &= \int_{B_\rho} \left(u(x) + \nabla u(x)\cdot y + \frac{1}{2}D^2u(x)y\cdot y + o(|y|^2)\right)\,dy.\end{aligned} \tag{1.3}$$

Also, by odd symmetry we see that

$$\int_{B_\rho} y_j\,dy = 0 \qquad \text{for all } j \in \{1,\dots,n\}$$

and

$$\int_{B_\rho} y_j\,y_k\,dy = 0 \qquad \text{for all } j \neq k \in \{1,\dots,n\}.$$

Consequently, we can write (1.3) as

$$\begin{aligned}\fint_{B_\rho(x)} u(y)\,dy &= u(x) + \frac{1}{2}\sum_{j=1}^{n} \partial^2_{jj}u(x) \fint_{B_\rho} y_j^2\,dy + o(\rho^2) \\ &= u(x) + \frac{1}{2n}\sum_{j=1}^{n} \partial^2_{jj}u(x) \fint_{B_\rho} |y|^2\,dy + o(\rho^2) \\ &= u(x) + \frac{\rho^2}{2(n+2)}\,\Delta u(x) + o(\rho^2)\end{aligned}$$

and therefore

$$-\Delta u(x) = 2(n+2)\lim_{\rho\searrow 0} \fint_{B_\rho(x)} \frac{u(x)-u(y)}{\rho^2}\,dy. \tag{1.4}$$

Similarly,

$$\begin{aligned}
\fint_{\partial B_\rho(x)} u(y)\, d\mathcal{H}^{n-1}(y) &= \fint_{\partial B_\rho} \Big(u(x) + \nabla u(x)\cdot y + \frac{1}{2}\, D^2 u(x) y \cdot y \\
&\quad + o(|y|^2) \Big)\, d\mathcal{H}^{n-1}(y) \\
&= u(x) + \frac{1}{2} \sum_{j=1}^{n} \partial^2_{jj} u(x) \fint_{\partial B_\rho} y_j^2 \, d\mathcal{H}^{n-1}(y) + o(\rho^2) \\
&= u(x) + \frac{1}{2n} \sum_{j=1}^{n} \partial^2_{jj} u(x) \fint_{\partial B_\rho} |y|^2 \, d\mathcal{H}^{n-1}(y) + o(\rho^2) \\
&= u(x) + \frac{\rho^2}{2n}\, \Delta u(x) + o(\rho^2)
\end{aligned}$$

and therefore

$$\begin{aligned}
-\Delta u(x) &= 2n \lim_{\rho \searrow 0} \fint_{\partial B_\rho(x)} \frac{u(x) - u(y)}{\rho^2}\, d\mathcal{H}^{n-1}(y) \\
&= 2n \lim_{\rho \searrow 0} \fint_{\partial B_\rho(x)} \frac{u(x) - u(y)}{|x-y|^2}\, d\mathcal{H}^{n-1}(y).
\end{aligned} \tag{1.5}$$

Once again, the factors $2(n + 2)$ and $2n$ in (1.4) and (1.5) are not important for our purposes, but the similarities between (1.2), (1.4) and (1.5) are evident and suggest that the fractional Laplacian is a suitably weighted average distributed in the whole of the space.

3. The classical Laplace operator is variational and stems from a Dirichlet energy of the form

$$\int |\nabla u(x)|^2 \, dx. \tag{1.6}$$

Similarly, the fractional Laplacian is variational and the corresponding energy is the Gagliardo-Slobodeckij-Sobolev seminorm

$$\iint \frac{|u(x) - u(y)|^2}{|x-y|^{n+2s}}\, dx\, dy. \tag{1.7}$$

The integral in (1.6) usually ranges in a "domain" $\Omega \subseteq \mathbb{R}^n$ which should be considered as the region of space where "action takes place", or, better to say the complement of the region in which no action takes place (that is, the domain Ω is the complement of the region $\mathbb{R}^n \setminus \Omega$, where the data of u are fixed). The fractional counterpart of this is to take as "natural domain" for (1.7) the complement (in $\mathbb{R}^{2n}$) of the set $(\mathbb{R}^n \setminus \Omega) \times (\mathbb{R}^n \setminus \Omega)$ where the data of $u(x) - u(y)$ are fixed, that is, it is common to integrate (1.7) over the "cross domain"

$$\begin{aligned} \mathcal{Q}_\Omega &:= (\Omega \times \Omega) \cup \big(\Omega \times (\mathbb{R}^n \setminus \Omega)\big) \cup \big((\mathbb{R}^n \setminus \Omega) \times \Omega\big) \\ &= \mathbb{R}^{2n} \setminus \big((\mathbb{R}^n \setminus \Omega) \times (\mathbb{R}^n \setminus \Omega)\big). \end{aligned}$$

4. Most importantly, the fractional Laplacian enjoys "elliptic" features that are similar to the ones of the classical Laplacian, e.g. in terms of maximum principle. The regularizing effects of the fractional Laplacian can be somewhat "guessed" from the singularity of the integral kernel in (1.1): indeed, on the one hand, to make sense of the integral in (1.1), one needs the function u to be "smooth enough" near x; on the other hand, and somehow conversely, if the integral in (1.1) is finite, the function u needs to have some regularity property near x, in order to compensate the singularity of the kernel.

Several classical and recent publications presented the fractional Laplacian from different perspectives. See in particular [5, 10, 24, 32, 34]. In our postmodern world some excellent online expositions of this topic have also become available, see in particular the very useful webpage https://www.ma.utexas.edu/mediawiki/index.php/Fractional_Laplacian.

We also recall that the fractional Laplacian can also be framed into the context of probability and harmonic analysis, thus leading to different possible approaches and several possible definitions, see [23], and it is also possible to provide a suitable setting in order to define the fractional Laplacian for functions with polynomial growth at infinity, see [14].

In spite of the extremely important similarities between the classical and the fractional Laplacian, several structural differences between these operators arise. See e.g. [1] for a collection of some of these basic differences. Some of these differences have also extremely deep consequences on some recent results in the theory of nonlocal equations, see https://www.ma.utexas.edu/mediawiki/index.php/List_of_results_that_are_fundamentally_different_to_the\discretionary-local_case.

In this note, we recall one of the basic differences between the classical and the fractional Laplacian, which has been recently discovered in [13] and which presents a source of interesting consequences. This difference deals with the so called "s-harmonic functions", which are the (rather surprising) counterpart of classical harmonic functions.

The parallelism between classical harmonic functions and s-harmonic functions lies in their definition, since u is said to be harmonic (respectively, s-harmonic) at x if $-\Delta u(x) = 0$ (respectively, if $(-\Delta)^s u(x) = 0$).

Already from the definition, a basic difference between the classical and the fractional case arises, since the definition of harmonic function at x only requires the function to be defined in an arbitrarily small neighborhood of x, while the definition of s-harmonic function requires the function to be globally defined in $\mathbb{R}^n$. This difference, which is somehow the counterpart of the structural differences between (1.2) on one side and (1.4) and (1.5) on the other side, turns out to be perhaps deeper than what may look at a first glance. As a matter of fact, the classical Laplacian is a very "rigid" operator, and for a function to be harmonic some very restrictive geometric conditions must hold (in particular, harmonic functions cannot have local minima). In sharp contrast with this fact, the fractional Laplacian is very flexible and the "oscillations of a function that come from far" can locally produce very significant contributions.

Probably, the most striking example of this phenomenon is that such far-away oscillations can make the fractional Laplacian of any function to almost vanish at a point, and in fact any given function, without any restriction on its geometric properties, can be approximated arbitrarily well by an s-harmonic function. In this sense, we have:

Theorem 1 ("All Functions Are Locally s-Harmonic up to a Small Error" [13]) *For any $\varepsilon > 0$ and any function $\bar{v} \in C^2(\overline{B_1})$, there exists v_ε such that*

$$\begin{cases} \|\bar{v} - v_\varepsilon\|_{C^2(B_1)} \leqslant \varepsilon, \\ (-\Delta)^s v_\varepsilon = 0 \text{ in } B_1. \end{cases}$$

A proof of this fact (in dimension 1 for the sake of simplicity) will be given in Sect. 3 (see the original paper [13] for the full details of the argument in any dimension).

We stress that the phenomenon described in Theorem 1 is very general, and it arises also for other nonlocal operators, independently from their possibly "elliptic" structure (for instance all functions are locally s-caloric, or s-hyperbolic, etc.), see [11].

It is interesting to remark that the proofs in [11, 13] are not "quantitative", in the sense that they are based on a contradiction argument, and the "shape" of the approximating s-harmonic (or s-caloric, or s-hyperbolic) function cannot be detected by our methods. On the other hand, for a very nice quantitative version of Theorem 1 see Theorem 1.4 in [29]. See also [30] for a quantitative approach to the parabolic case and [18] for related results (and, of course, quantitative proofs are harder and technically more advanced than the one that we present here). In addition, results similar to Theorem 1 hold true for nonlocal operators with memory, see [4].

Theorem 1 possesses some simple, but quite interesting consequences. In the forthcoming Sect. 2 we present a few of them, related to

1. The fractional Maximum Principle and Harnack Inequality;
2. The classification of stable solutions for fractional equations;
3. The diffusive strategy of biological populations.

2 Applications of Theorem 1

2.1 *The Fractional Maximum Principle and Harnack Inequality*

One of the main features of the classical Laplace operator is that it enjoys the Maximum Principle. For instance, as well known, it holds that:

Theorem 2 *Let u be a harmonic and nonnegative function in B_1. If $u(x_0) = 0$ for some $x_0 \in B_1$, then u is necessarily constantly equal to 0 in B_1.*

A classical quantitative version of Theorem 2 was given by Axel von Harnack and can be stated as follows:

Theorem 3 *If u is harmonic in B_1 and nonnegative in B_1, then, for every $r \in (0, 1)$,*

$$\sup_{B_r} u \leqslant C_r \inf_{B_r} u,$$

for some $C_r > 0$ depending on n and r.

The original manuscript by von Harnack is available at https://ia902306.us.archive.org/9/items/vorlesunganwend00weierich/vorlesunganwend00weierich.pdf.

The fractional counterpart of Theorem 3 goes as follows:

Theorem 4 *If u is s-harmonic in B_1 and nonnegative in the whole of $\mathbb{R}^n$, then, for every $r \in (0, 1)$,*

$$\sup_{B_r} u \leqslant C_r \inf_{B_r} u,$$

for some $C_r > 0$ depending on n and r.

See [2, 6, 19, 21, 22, 35] and the references therein for a detailed study of the fractional Harnack Inequality. Of course, an important structural difference between Theorems 3 and 4 (besides the s-harmonicity versus the classical harmonicity) is the fact that in Theorem 4 one requires a global condition on the sign of the solution. Interestingly, if in Theorem 4 one replaces the assumption that u is nonnegative in the full space with the assumption that u is nonnegative just in the unit ball, then the result turns out to be false, as described by the following example:

Theorem 5 *There exists a bounded function u which is s-harmonic in B_1, nonnegative in B_1, not identically 0 in B_1, but such that*

$$\inf_{B_{1/2}} u = 0.$$

Theorem 5 suggests that some care has to be taken when dealing with Maximum Principles and oscillation results in the fractional case, and in fact the nonlocal character of the operator requires global conditions for this type of results to hold, in virtue of the contributions "coming from far away".

A proof of Theorem 5 can be obtained directly from Theorem 1. Indeed, we take $n = 1$, $\bar{v}(x) := x^2$ and $\varepsilon := \frac{1}{16}$. Then, Theorem 1 provides a function v which is s-harmonic in $(-1, 1)$ and such that

$$\|\bar{v} - v\|_{L^\infty((-1,1))} \leqslant \|\bar{v} - v\|_{C^2((-1,1))} \leqslant \frac{1}{16}.$$

In particular, if $|x| \geqslant \frac{1}{2}$,

$$v(x) \geqslant \bar{v}(x) - \frac{1}{16} = |x|^2 - \frac{1}{16} \geqslant \left(\frac{1}{2}\right)^2 - \frac{1}{16} = \frac{3}{16},$$

while

$$v(0) \leqslant \bar{v}(0) + \frac{1}{16} = \frac{1}{16}.$$

Accordingly,

$$\inf_{(-1,1)} v \leqslant \frac{1}{16} < \frac{3}{16} \leqslant \inf_{(-1,1)\setminus(-1/2,1/2)} v,$$

which gives that

$$\inf_{(-1,1)} v = \min_{[-1/2,1/2]} v =: \iota.$$

Then the function $u := v - \iota$ satisfies the thesis of Theorem 5, as desired.

For different approaches to the counterexamples to the local Harnack Inequality in the fractional setting see [3, 20] and also[1] Chapter 2.3 in [5].

[1] We take this opportunity to amend a typo in Theorem 2.3.1 of [5], where $\inf_{B_1} u$ has to be replaced by $\inf_{B_{1/2}} u$.

2.2 The Classification of Stable Solutions for Fractional Equations

In the Calculus of Variations[2] literature, a solution u is called "stable" if it is the critical point of an energy functional whose second variation is nonnegative definite at u. For instance, local minimizers of the energy are stable solutions, and it is in fact often convenient to study stable solutions since the stability class is often preserved under suitable limit procedures and it is sometimes technically easier (or at least less difficult) to prove that a solution is stable rather than deciding whether or not it is minimal.

We refer to the very nice monograph [16] for a throughout discussion of the notion of stability and for many related results.

A classical result in the framework of stable solutions of elliptic equations was obtained independently by Richard Casten and Charles Holland, on the one side, and Hiroshi Matano, on the other side, and it deals with the classification of stable solutions with Neumann data. A paradigmatic result in this case can be stated as follows:

Theorem 6 ([9, 26]) *Let $\Omega \subset \mathbb{R}^n$ be a bounded and convex domain with smooth boundary.*

Suppose that u is a smooth solution of

$$\begin{cases} -\Delta u(x) + f(u(x)) = 0 & \text{for any } x \in \Omega \\ \dfrac{\partial u}{\partial \nu}(x) = 0 & \text{for any } x \in \partial\Omega, \end{cases} \tag{2.1}$$

for some smooth function f, where ν denotes the (external) unit normal of Ω.

Assume also that u is stable, namely

$$\int_{\Omega} |\nabla\varphi(x)|^2 + f'(u(x))\,|\varphi(x)|^2\,dx \geqslant 0, \tag{2.2}$$

for any $\varphi \in H^1(\Omega)$.

Then, u is necessarily constant.

We remark that

$$\text{when } f \text{ vanishes identically then (2.2) is automatically satisfied.} \tag{2.3}$$

[2] Notice that the notion of "stability" differs from one scientific community to another. In particular, the notion of stability that we treat here does not agree with that in Dynamical Systems or Algebraic Geometry.

It is interesting to observe that, with respect to Theorem 6, the fractional case behaves very differently, and nonconstant stable solutions with Neumann data in convex domains do exist, according to the following result:

Theorem 7 ([15]) *Let $s \in (0,1)$. There exist an open interval $I \subset \mathbb{R}$ and a nonconstant function u such that*

$$(-\Delta)^s u = 0 \ \ in \ \ I, \tag{2.4}$$

$$\lim_{x \to x_0 \in \partial I} \frac{u(x) - u(x_0)}{x - x_0} = 0 \tag{2.5}$$

$$and\, u' = 0 \ \ on \ \ \partial I. \tag{2.6}$$

We observe that (2.4) is the natural fractional counterpart of (2.1) (with $f := 0$, and (2.3) guarantees a stability condition). Of course, an interval is a (onedimensional) convex set, hence the geometric setting of Theorem 6 is respected in Theorem 7. Also, formula (2.6) can be seen as a classical Neumann condition, while formula (2.5) can be seen as a fractional Neumann condition (say, of order s). Condition (2.5) is indeed quite exploited as a natural boundary condition in fractional problems, and it is compatible with the boundary regularity theory and with the sliding methods, see [17, 28] (for another notion of fractional Neumann condition see [12]).

In this sense, Theorem 7 can be considered as a "counterexample" for the fractional analogue of Theorem 6 to hold. The construction of Theorem 7 is in fact very general. It is based on Theorem 1 and provides a series of rather "arbitrary" counterexamples, see Section 1.7 in [15] for additional details.

It has to be pointed out, however, that results similar to the original ones in [9, 26] hold true for a different type of fractional operator (the so-called "spectral" fractional Laplacian, see [31]). In particular, classification results for stable solutions of nonlocal operators which can be seen as the fractional counterpart of those in [9, 26] have been given in Sections 1.4–1.6 in [15]. This fact shows the very intriguing phenomenon, according to which "little" modifications in the fractional settings do produce rather different results, which are sometimes in agreement with the classical case, and sometimes not.

2.3 *The Diffusive Strategy of Biological Populations*

A classical problem in biomathematics consists in studying the evolution of a biological species with density $u = u(x,t)$ in $B_1 \ni x$, with prescribed boundary or external conditions. In this framework, the so-called logistic equations is based on the ansatz that the state of the population is due to three well distinguishable features:

- The population diffuses according to a stochastic motion;
- For small density, the population grows more or less linearly, thanks to some resources $\rho = \rho(x) > 0$;
- When the density overcomes a critical threshold σ/μ, for some $\mu = \mu(x) > 0$, the population unfortunately dies (roughly speaking, because "there is no food for everybody").

When the diffusion term is led by the standard Brownian motion, the logistic equation that we describe takes the form

$$\partial_t u = \Delta u + (\sigma - \mu u)\, u \qquad \text{in } B_1 \times (0, T), \tag{2.7}$$

for some $T > 0$. In particular, the study of the steady states of (2.7) leads to the equation

$$-\Delta u = (\sigma - \mu u)\, u \qquad \text{in } B_1. \tag{2.8}$$

On the other hand, recent experiments have shown that several predators do not follow standard diffusion processes, but rather discontinuous processes with jumps whose distribution may exhibit a long (e.g. with a polynomial tail), see e.g. [36]. This fact, that may seem surprising, has indeed a sound motivation: for a predator it makes little sense to move randomly looking for prey, since, after a first attack, the other possible targets will rapidly escape from the dangerous area—conversely, a strategy of "hit and run", based on quick hunts after long excursions, is more reasonable to be efficient and ensure more food to the predator.

In this sense, a natural nonlocal variation of (2.8) to be taken into account is the fractional logistic equation

$$(-\Delta)^s u = (\sigma - \mu u)\, u \qquad \text{in } B_1, \tag{2.9}$$

with $s \in (0, 1)$, see e.g. [7, 8, 25, 27] and the references therein. Interestingly, different species in nature seem to exhibit different values of the fractional parameter s, probably due to different environmental conditions and different morphological structures and it is an intriguing problem to understand what "the optimal exponent s" should be in concrete circumstances, see [33].

Another interesting special feature offered by nonlocal diffusion is the possibility for nonlocal populations to efficiently plan their distribution in order to consume all (or almost all) the given resources in a certain "strategic region". That is, if the region of interest for the population is, say, the ball B_1, the species can artificially and appropriately settle its distribution outside B_1, in order to satisfy in B_1 a logistic equation as that in (2.9), for a resource that is arbitrarily close to the original one. The precise statement of this result is the following:

Theorem 8 ([7, 25]) *Assume that* $\sigma,\ \mu \in C^2(\overline{B_1})$, *with*

$$\inf_{B_1} \sigma > 0 \quad \textit{and} \quad \inf_{B_1} \mu > 0.$$

Then, for any $\varepsilon > 0$ *there exist* u_ε *and* σ_ε *such that*

$$\begin{cases} \|\sigma - \sigma_\varepsilon\|_{C^2(B_1)} \leqslant \varepsilon, \\ u_\varepsilon \geqslant \sigma_\varepsilon/\mu \quad \textit{in } B_1, \\ (-\Delta)^s u_\varepsilon = (\sigma_\varepsilon - \mu u_\varepsilon)\, u_\varepsilon \ \textit{in } B_1. \end{cases}$$

Once again, a proof of Theorem 8 may be performed by exploiting Theorem 1, see Section 7 in [7].

3 Proof of Theorem 1

For simplicity, we focus on the one-dimensional case: the general case follows by technical modifications and can be found in the original article [13].

The core of the proof is to show that the derivatives of s-harmonic functions have "maximal span" as a linear space (and we stress that this is not true for harmonic functions, since the second derivatives of harmonic functions satisfy a linear prescription).

We consider the set

$$\begin{aligned} \mathscr{V} := \big\{ & h : \mathbb{R} \to \mathbb{R} \text{ s.t. } h \text{ is smooth and} \\ & s\text{-harmonic in someneighborhood of the origin} \big\}. \end{aligned} \tag{3.1}$$

Notice that $\mathscr{V}$ has a linear space structure, namely if h_1 is s-harmonic in some open set V_1 containing the origin and h_2 is s-harmonic in some open set V_2 containing the origin, then, for any $\lambda_1,\ \lambda_2 \in \mathbb{R}$, we have that $h_3 := \lambda_1 h_1 + \lambda_2 h_2$ is s-harmonic in the open set $V_3 := V_1 \cap V_2 \ni 0$.

Then, given $J \in \mathbb{N}$, we define

$$\mathscr{V}_J := \Big\{ \big(h(0),\ h'(0),\ \ldots,\ h^{(J)}(0) \big) \text{ with } h \in \mathscr{V} \Big\}. \tag{3.2}$$

As customary, here $h^{(J)}$ denotes the Jth derivative of the function h. In this way, we have that $\mathscr{V}_J$ is a linear subspace of $\mathbb{R}^{J+1}$ (roughly speaking, each element of $\mathscr{V}_J$ is a $(J+1)$-dimensional array containing the first J derivatives of a locally s-harmonic function).

We claim that

$$\mathscr{V}_J = \mathbb{R}^{J+1} \tag{3.3}$$

For this, we argue by contradiction and we suppose that $\mathscr{V}_J$ is a linear subspace strictly smaller than $\mathbb{R}^{J+1}$. That is, $\mathscr{V}_J$ lies inside a J-dimensional hyperplane, say with normal ν. Namely, there exists

$$\nu = (\nu_0, \ldots, \nu_J) \in \mathbb{R}^{J+1} \text{ with } |\nu| = 1 \tag{3.4}$$

such that

$$\mathscr{V}_J \subseteq \big\{X = (X_0, \ldots, X_J) \in \mathbb{R}^{J+1} \text{ s.t. } \nu \cdot X = 0\big\} \tag{3.5}$$

Now, for any $t > 0$, we define

$$h_t(x) := (x+t)^s_+.$$

It is known that h_t is s-harmonic in $(-t, +\infty)$ (see e.g. Chapter 2.4 in [5] for an elementary proof). Consequently, $h_t \in \mathscr{V}$ and then

$$X_t := \big(h_t(0),\ \ldots,\ h_t^{(J)}(0)\big) \in \mathscr{V}_J.$$

As a result, by (3.5),

$$0 = \nu \cdot X_t = \sum_{j=0}^{J} \nu_j h_t^{(j)}(0) = \sum_{j=0}^{J} \mu_{s,j}\, t^{s-j}, \tag{3.6}$$

where

$$\mu_{s,j} := \nu_j \prod_{i=0}^{j-1} (s-i). \tag{3.7}$$

Hence, multiplying the identity in (3.6) by t^{J-s}, for any $t > 0$, it holds that

$$\sum_{k=0}^{J} \mu_{s,J-k}\, t^k = 0,$$

which, by the Identity Principle for Polynomials, implies that $\mu_{s,0} = \cdots = \mu_{s,J} = 0$ and accordingly[3] from (3.7) we get that $\nu_0 = \cdots = \nu_J = 0$. This is in contradiction with (3.4) and so the proof of (3.3) is complete.

[3] We stress that here it is crucially used the fact that s is not an integer.

Now, the proof of Theorem 1 follows by approximation and scaling. Given $\bar{v} \in C^2(\overline{B_1})$ and $\varepsilon \in (0,1)$, in view of the Stone-Weierstrass Theorem we take a polynomial $\mathscr{P}_\varepsilon$ such that

$$\|\bar{v} - \mathscr{P}_\varepsilon\|_{C^2(B_1)} \leqslant \frac{\varepsilon}{2}. \tag{3.8}$$

We write

$$\mathscr{P}_\varepsilon(x) = \sum_{j=0}^{N_\varepsilon} c_{j,\varepsilon} x^j = \sum_{j=0}^{N_\varepsilon} m_{j,\varepsilon}(x),$$

for some $N_\varepsilon \in \mathbb{N}$ and some $c_{1,\varepsilon}, \dots, c_{N_\varepsilon,\varepsilon} \in \mathbb{R}$, where

$$m_{j,\varepsilon}(x) := c_{j,\varepsilon} x^j. \tag{3.9}$$

Without loss of generality, by possibly adding zero coefficients in the representation above, we can suppose that

$$N_\varepsilon \geqslant 3. \tag{3.10}$$

We set

$$C_\varepsilon := \max_{j \in \{0,\dots,N_\varepsilon\}} |c_{j,\varepsilon}|.$$

For any $j \in \{0, \dots, N_\varepsilon\}$, we let $H_{j,\varepsilon} : \mathbb{R} \to \mathbb{R}$ be a function which is s-harmonic in a neighborhood of the origin and such that, for any $i \in \{0, \dots, N_\varepsilon\}$ it holds that

$$H_{j,\varepsilon}^{(i)}(0) = \begin{cases} c_{j,\varepsilon}\, j! & \text{if } i = j, \\ 0 & \text{otherwise.} \end{cases} \tag{3.11}$$

Once again, $H_{j,\varepsilon}^{(i)}$ denotes here the ith derivative of $H_{j,\varepsilon}$. We stress that the existence of $H_{j,\varepsilon}$ is a consequence of (3.3). We also set

$$\begin{aligned} r_{j,\varepsilon} &:= \frac{\varepsilon}{10 N_\varepsilon^2 \left(1 + \sup\limits_{x \in (-1,1)} \left|H_{j,\varepsilon}^{(N_\varepsilon+1)}(x)\right|\right)} \in (0,1) \\ \text{and } \mathscr{H}_{j,\varepsilon}(x) &:= r_{j,\varepsilon}^{-j}\, H_{j,\varepsilon}(r_{j,\varepsilon} x). \end{aligned} \tag{3.12}$$

We remark that, for any $i, j \in \{0, \dots, N_\varepsilon\}$,

$$\mathscr{H}_{j,\varepsilon}^{(i)}(0) = \begin{cases} c_{j,\varepsilon}\, j! & \text{if } i = j, \\ 0 & \text{otherwise,} \end{cases}$$

thanks to (3.11). Therefore, in view of (3.9), the function

$$\mathscr{D}_{j,\varepsilon}(x) := \mathscr{H}_{j,\varepsilon}(x) - c_{j,\varepsilon}x^j = \mathscr{H}_{j,\varepsilon}(x) - m_{j,\varepsilon}(x) \tag{3.13}$$

satisfies

$$\mathscr{D}_{j,\varepsilon}^{(i)}(0) = 0 \text{ for all } i \in \{0, \dots, N_\varepsilon\}. \tag{3.14}$$

In addition, for any $x \in (-1, 1)$ and any $j \in \{0, \dots, N_\varepsilon\}$,

$$\begin{aligned}
\left|\mathscr{D}_{j,\varepsilon}^{(N_\varepsilon+1)}(x)\right| &= \left|\mathscr{H}_{j,\varepsilon}^{(N_\varepsilon+1)}(x)\right| \\
&\leqslant r_{j,\varepsilon}^{N_\varepsilon+1-j}\left|H_{j,\varepsilon}^{(N_\varepsilon+1)}(r_{j,\varepsilon}x)\right| \\
&\leqslant r_{j,\varepsilon} \sup_{(-1,1)} \left|H_{j,\varepsilon}^{(N_\varepsilon+1)}\right| \\
&\leqslant \frac{\varepsilon}{2N_\varepsilon^2},
\end{aligned}$$

thanks to (3.12). This, (3.14) and a Taylor expansion give that, for any $x \in (-1, 1)$ and any $i, j \in \{0, \dots, N_\varepsilon\}$,

$$\left|\mathscr{D}_{j,\varepsilon}^{(i)}(x)\right| \leqslant \sup_{(-1,1)} \left|\mathscr{D}_{j,\varepsilon}^{(N_\varepsilon+1)}\right| \leqslant \frac{\varepsilon}{10\, N_\varepsilon^2}.$$

Hence, recalling (3.10)

$$\sum_{j=0}^{N_\varepsilon} \|\mathscr{D}_{j,\varepsilon}\|_{C^2(-1,1)} \leqslant \frac{\varepsilon}{2}.$$

So, we define

$$v_\varepsilon := \sum_{j=0}^{N_\varepsilon} \mathscr{H}_{j,\varepsilon}.$$

We have that v_ε is s-harmonic in $(-1, 1)$ and, recalling (3.8) and (3.13),

$$\begin{aligned}
\|\bar{v} - v_\varepsilon\|_{C^2(-1,1)} &\leqslant \|\bar{v} - \mathscr{P}_\varepsilon\|_{C^2(-1,1)} + \|\mathscr{P}_\varepsilon - v_\varepsilon\|_{C^2(-1,1)} \\
&\leqslant \frac{\varepsilon}{2} + \left\|\sum_{j=0}^{N_\varepsilon} \left(m_{j,\varepsilon} - \mathscr{H}_{j,\varepsilon}\right)\right\|_{C^2(-1,1)}
\end{aligned}$$

$$
\begin{aligned}
&\leqslant \frac{\varepsilon}{2} + \sum_{j=0}^{N_\varepsilon} \|\mathscr{D}_{j,\varepsilon}\|_{C^2(-1,1)} \\
&\leqslant \varepsilon.
\end{aligned}
$$

This establishes Theorem 1 in this setting.

Acknowledgements Enrico Valdinoci is supported by the Istituto Nazionale di Alta Matematica and the Australian Research Council Discovery Project *N.E.W.* "Nonlocal Equations at Work".

References

1. N. Abatangelo, E. Valdinoci, *Getting Acquainted with the Fractional Laplacian*. Springer INdAM Series (Springer, Cham, to appear)
2. R.F. Bass, D.A. Levin, Harnack inequalities for jump processes. Potential Anal. **17**(4), 375–388 (2002). https://doi.org/10.1023/A:1016378210944. MR1918242
3. K. Bogdan, P. Sztonyk, Harnack's inequality for stable Lévy processes. Potential Anal. **22**(2), 133–150 (2005). https://doi.org/10.1007/s11118-004-0590-x. MR2137058
4. C. Bucur, Local density of Caputo-stationary functions in the space of smooth functions. ESAIM Control Optim. Calc. Var. **23**(4), 1361–1380 (2017)
5. C. Bucur, E. Valdinoci, *Nonlocal Diffusion and Applications*. Lecture Notes of the Unione Matematica Italiana, vol. 20 (Springer, Cham; Unione Matematica Italiana, Bologna, 2016). MR3469920
6. L. Caffarelli, L. Silvestre, Regularity theory for fully nonlinear integro-differential equations. Commun. Pure Appl. Math. **62**(5), 597–638 (2009). https://doi.org/10.1002/cpa.20274. MR2494809
7. L. Caffarelli, S. Dipierro, E. Valdinoci, A logistic equation with nonlocal interactions. Kinet. Relat. Models **10**(1), 141–170 (2017). https://doi.org/10.3934/krm.2017006. MR3579567
8. G. Carboni, D. Mugnai, On some fractional equations with convex-concave and logistic-type nonlinearities. J. Differ. Equ. **262**(3), 2393–2413 (2017). https://doi.org/10.1016/j.jde.2016.10.045. MR3582231
9. R.G. Casten, C.J. Holland, Instability results for reaction diffusion equations with Neumann boundary conditions. J. Differ. Equ. **27**(2), 266–273 (1978). https://doi.org/10.1016/0022-0396(78)90033-5. MR480282
10. E. Di Nezza, G. Palatucci, E. Valdinoci, Hitchhiker's guide to the fractional Sobolev spaces. Bull. Sci. Math. **136**(5), 521–573 (2012). https://doi.org/10.1016/j.bulsci.2011.12.004. MR2944369
11. S. Dipierro, O. Savin, E. Valdinoci, Local approximation of arbitrary functions by solutions of nonlocal equations. J. Geom. Anal. (to appear)
12. S. Dipierro, X. Ros-Oton, E. Valdinoci, Nonlocal problems with Neumann boundary conditions. Rev. Mat. Iberoam. **33**(2), 377–416 (2017). https://doi.org/10.4171/RMI/942. MR3651008
13. S. Dipierro, O. Savin, E. Valdinoci, All functions are locally s-harmonic up to a small error. J. Eur. Math. Soc. **19**(4), 957–966 (2017). https://doi.org/10.4171/JEMS/684. MR3626547
14. S. Dipierro, O. Savin, E. Valdinoci, Definition of fractional Laplacian for functions with polynomial growth. Rev. Mat. Iberoam. (to appear)
15. S. Dipierro, N. Soave, E. Valdinoci, On stable solutions of boundary reaction-diffusion equations and applications to nonlocal problems with Neumann data. Indiana Univ. Math. J. **67**(1), 429–469 (2018)

16. L. Dupaigne, *Stable Solutions of Elliptic Partial Differential Equations*. Chapman and Hall/CRC Monographs and Surveys in Pure and Applied Mathematics, vol. 143 (Chapman and Hall/CRC, Boca Raton, 2011). MR2779463
17. M.M. Fall, S. Jarohs, Overdetermined problems with fractional Laplacian. ESAIM Control Optim. Calc. Var. **21**(4), 924–938 (2015). https://doi.org/10.1051/cocv/2014048. MR3395749
18. T. Ghosh, M. Salo, G. Uhlmann, The Calderón problem for the fractional Schrödinger equation, arXiv e-prints (2016). Available at 1609.09248
19. M. Kaßmann, *Harnack-Ungleichungen für nichtlokale Differentialoperatoren und Dirichlet-Formen*. Bonner Mathematische Schriften [Bonn Mathematical Publications], vol. 336 (Universität Bonn, Mathematisches Institut, Bonn, 2001, in German). Dissertation, Rheinische Friedrich-Wilhelms-Universität Bonn, Bonn, 2000. MR1941020
20. M. Kaßmann, The classical Harnack inequality fails for non-local operators. Preprint SFB 611, Sonderforschungsbereich Singuläre Phänomene und Skalierung in Mathematischen Modellen, No. 360 (2007)
21. M. Kaßmann, A new formulation of Harnack's inequality for nonlocal operators. C. R. Math. Acad. Sci. Paris **349**(11–12), 637–640 (2011). https://doi.org/10.1016/j.crma.2011.04.014 (English, with English and French summaries). MR2817382
22. M. Kaßmann, M. Rang, R.W. Schwab, Integro-differential equations with nonlinear directional dependence. Indiana Univ. Math. J. **63**(5), 1467–1498 (2014). https://doi.org/10.1512/iumj.2014.63.5394. MR3283558
23. M. Kwaśnicki, Ten equivalent definitions of the fractional Laplace operator. Fract. Calc. Appl. Anal. **20**(1), 7–51 (2017). https://doi.org/10.1515/fca-2017-0002. MR3613319
24. N.S. Landkof, *Foundations of Modern Potential Theory* (Springer, New York, 1972). Translated from the Russian by A.P. Doohovskoy; Die Grundlehren der mathematischen Wissenschaften, Band 180. MR0350027
25. A. Massaccesi, E. Valdinoci, Is a nonlocal diffusion strategy convenient for biological populations in competition? J. Math. Biol. **74**(1–2), 113–147 (2017). https://doi.org/10.1007/s00285-016-1019-z. MR3590678
26. H. Matano, Asymptotic behavior and stability of solutions of semilinear diffusion equations. Publ. Res. Inst. Math. Sci. **15**(2), 401–454 (1979). https://doi.org/10.2977/prims/1195188180. MR555661
27. E. Montefusco, B. Pellacci, G. Verzini, Fractional diffusion with Neumann boundary conditions: the logistic equation. Discrete Contin. Dyn. Syst. Ser. B **18**(8), 2175–2202 (2013). https://doi.org/10.3934/dcdsb.2013.18.2175. MR3082317
28. X. Ros-Oton, J. Serra, The Dirichlet problem for the fractional Laplacian: regularity up to the boundary. J. Math. Pures Appl. (9) **101**(3), 275–302 (2014). https://doi.org/10.1016/j.matpur.2013.06.003 (English, with English and French summaries). MR3168912
29. A. Rüland, M. Salo, The fractional Calderón problem: low regularity and stability, ArXiv e-prints (2017). Available at 1708.06294
30. A. Rüland, M. Salo, Quantitative approximation properties for the fractional heat equation, arXiv e-prints (2017). Available at 1708.06300
31. R. Servadei, E. Valdinoci, On the spectrum of two different fractional operators. Proc. R. Soc. Edinb. Sect. A **144**(4), 831–855 (2014). https://doi.org/10.1017/S0308210512001783. MR3233760
32. L.E. Silvestre, *Regularity of the Obstacle Problem for a Fractional Power of the Laplace Operator* (ProQuest LLC, Ann Arbor, 2005). Thesis (Ph.D.)–The University of Texas at Austin. MR2707618
33. J. Sprekels, E. Valdinoci, A new type of identification problems: optimizing the fractional order in a nonlocal evolution equation. SIAM J. Control Optim. **55**(1), 70–93 (2017). https://doi.org/10.1137/16M105575X. MR3590646
34. E.M. Stein, *Singular Integrals and Differentiability Properties of Functions*. Princeton Mathematical Series, No. 30 (Princeton University Press, Princeton, 1970). MR0290095

35. P.R. Stinga, J.L. Torrea, Extension problem and Harnack's inequality for some fractional operators. Commun. Partial Differ. Equ. **35**(11), 2092–2122 (2010). https://doi.org/10.1080/03605301003735680. MR2754080
36. G.M. Viswanathan, V. Afanasyev, S.V. Buldyrev, E.J. Murphy, P.A. Prince, H.E. Stanley, Lévy flight search patterns of wandering albatrosses. Nature **381**, 413–415 (1996). https://doi.org/10.1038/381413a0

LECTURE NOTES IN MATHEMATICS Springer

Editorial Policy

1. Lecture Notes aim to report new developments in all areas of mathematics and their applications – quickly, informally and at a high level. Mathematical texts analysing new developments in modelling and numerical simulation are welcome.

   Manuscripts should be reasonably self-contained and rounded off. Thus they may, and often will, present not only results of the author but also related work by other people. They may be based on specialised lecture courses. Furthermore, the manuscripts should provide sufficient motivation, examples and applications. This clearly distinguishes Lecture Notes from journal articles or technical reports which normally are very concise. Articles intended for a journal but too long to be accepted by most journals, usually do not have this "lecture notes" character. For similar reasons it is unusual for doctoral theses to be accepted for the Lecture Notes series, though habilitation theses may be appropriate.

2. Besides monographs, multi-author manuscripts resulting from SUMMER SCHOOLS or similar INTENSIVE COURSES are welcome, provided their objective was held to present an active mathematical topic to an audience at the beginning or intermediate graduate level (a list of participants should be provided).

   The resulting manuscript should not be just a collection of course notes, but should require advance planning and coordination among the main lecturers. The subject matter should dictate the structure of the book. This structure should be motivated and explained in a scientific introduction, and the notation, references, index and formulation of results should be, if possible, unified by the editors. Each contribution should have an abstract and an introduction referring to the other contributions. In other words, more preparatory work must go into a multi-authored volume than simply assembling a disparate collection of papers, communicated at the event.

3. Manuscripts should be submitted either online at www.editorialmanager.com/lnm to Springer's mathematics editorial in Heidelberg, or electronically to one of the series editors. Authors should be aware that incomplete or insufficiently close-to-final manuscripts almost always result in longer refereeing times and nevertheless unclear referees' recommendations, making further refereeing of a final draft necessary. The strict minimum amount of material that will be considered should include a detailed outline describing the planned contents of each chapter, a bibliography and several sample chapters. Parallel submission of a manuscript to another publisher while under consideration for LNM is not acceptable and can lead to rejection.

4. In general, **monographs** will be sent out to at least 2 external referees for evaluation.

   A final decision to publish can be made only on the basis of the complete manuscript, however a refereeing process leading to a preliminary decision can be based on a pre-final or incomplete manuscript.

   Volume Editors of **multi-author works** are expected to arrange for the refereeing, to the usual scientific standards, of the individual contributions. If the resulting reports can be forwarded to the LNM Editorial Board, this is very helpful. If no reports are forwarded or if other questions remain unclear in respect of homogeneity etc, the series editors may wish to consult external referees for an overall evaluation of the volume.

5. Manuscripts should in general be submitted in English. Final manuscripts should contain at least 100 pages of mathematical text and should always include

   - a table of contents;
   - an informative introduction, with adequate motivation and perhaps some historical remarks: it should be accessible to a reader not intimately familiar with the topic treated;
   - a subject index: as a rule this is genuinely helpful for the reader.
   - For evaluation purposes, manuscripts should be submitted as pdf files.

6. Careful preparation of the manuscripts will help keep production time short besides ensuring satisfactory appearance of the finished book in print and online. After acceptance of the manuscript authors will be asked to prepare the final LaTeX source files (see LaTeX templates online: https://www.springer.com/gb/authors-editors/book-authors-editors/manuscriptpreparation/5636) plus the corresponding pdf- or zipped ps-file. The LaTeX source files are essential for producing the full-text online version of the book, see http://link.springer.com/bookseries/304 for the existing online volumes of LNM). The technical production of a Lecture Notes volume takes approximately 12 weeks. Additional instructions, if necessary, are available on request from lnm@springer.com.

7. Authors receive a total of 30 free copies of their volume and free access to their book on SpringerLink, but no royalties. They are entitled to a discount of 33.3 % on the price of Springer books purchased for their personal use, if ordering directly from Springer.

8. Commitment to publish is made by a *Publishing Agreement*; contributing authors of multiauthor books are requested to sign a *Consent to Publish form.* Springer-Verlag registers the copyright for each volume. Authors are free to reuse material contained in their LNM volumes in later publications: a brief written (or e-mail) request for formal permission is sufficient.

Addresses:

Professor Jean-Michel Morel, CMLA, École Normale Supérieure de Cachan, France
E-mail: moreljeanmichel@gmail.com

Professor Bernard Teissier, Equipe Géométrie et Dynamique,
Institut de Mathématiques de Jussieu – Paris Rive Gauche, Paris, France
E-mail: bernard.teissier@imj-prg.fr

Springer: Ute McCrory, Mathematics, Heidelberg, Germany,
E-mail: lnm@springer.com

Printed in the United States
By Bookmasters